NEW

Co-ordinated SCIENCE

Biology

TEACHER'S GUIDE

Brian Beckett

George Bethell

Barrie Crowther

RoseMarie Gallagher

Chris Goodwin

Paul Ingram

Stephen Pople

Oxford University Press

Oxford University Press, Walton Street, Oxford OX2 6DP

Oxford New York
Athens Auckland Bangkok Bogota Bombay
Buenos Aires Calcutta Cape Town Dar es Salaam Delhi
Florence Hong Kong Istanbul Karachi
Kuala Lumpur Madras Madrid Melbourne
Mexico City Nairobi Paris Singapore
Taipei Tokyo Toronto

and associated companies in
Berlin Ibadan

© OUP, George Bethell, Brian Beckett,
Chris Goodwin 1996

ISBN 019 914 2963

Produced by **AMR** Ltd

Printed in Great Britain by St Edmundsbury Press Ltd,
Bury St Edmunds, Suffolk

Introduction

Co-ordinated Science from Oxford is an ideal resource for use with KS4 GCSE science courses. It consists of three *Students' Books* (Physics, Chemistry, and Biology) each with an accompanying *Teacher's Guide*.

PART A contains:

- a concise series of notes drawing together links between the separate science subjects - ideal for minimizing overlap
- cross-referencing to related topics in the other *Students' Books*
- a list of relevant Activities, Investigations, and Assessment Tasks across the resource as a whole

PART B contains:

- 48 photocopiable Activities sheets for general use throughout the course

PART C contains:

- 12 Investigations and 6 Assessment Tasks which have been constructed to provide more formal opportunities within the area of Science 1

A list of Risk Assessments is given.

Contents

Risk Assessments

The following Risk Assessments for the various activities in this Teacher's Guide are offered for guidance only. Whilst they have been prepared in accordance with the currently existing guidelines for good practice, they should be viewed only as contributing to safe experimental work in school science laboratories and not taken as definitive. Teachers are advised that safety procedures recommended for good practice are continually being revised and that these changes should be monitored and incorporated accordingly. There may also be special local requirements placed on an institution which will be outside the scope of the guidelines offered here.

Copies of these HAZARD WARNINGS are repeated on the individual worksheets.

ACTIVITIES

A2 (p41) HAZARD WARNING
Copper sulphate is harmful when swallowed. It may also be irritating to eyes and skin. Potassium manganate VII is harmful if swallowed. AVOID SKIN CONTACT. WEAR EYE PROTECTION.

A3 (p42) HAZARD WARNING
Scalpels or razors are sharp, handle with care.

A4 (p43) HAZARD WARNING
Razor blades are sharp, handle with care.

A7 (p46) HAZARD WARNING
Ethanol is highly flammable. KEEP AWAY from naked flame. WEAR EYE PROTECTION.

A9 (p48) HAZARD WARNING
Ethanol is highly flammable. KEEP AWAY from naked flame. WEAR EYE PROTECTION.

A10 (p49) HAZARD WARNING
Ethanol is highly flammable. KEEP AWAY from naked flame. Soda lime (sodium hydroxide and calcium hydroxide) is CORROSIVE and can cause severe burns; also dangerous to eyes and skin. AVOID SKIN CONTACT. AVOID CONTACT WITH WATER. WEAR EYE PROTECTION.

A11 (p50) HAZARD WARNING
Razor blades are sharp, handle with care.

A12 (p51) HAZARD WARNING
Razor blades are sharp, handle with care.

A16 (p55) HAZARD WARNING
Hydrochloric acid is harmful. Avoid skin contact. WEAR EYE PROTECTION. When breaking glass rods WEAR EYE PROTECTION and use a safety screen.

A18 (p57) HEALTH CHECK
Check with your teacher that you are able to participate in this activity.

A19 (p58) HAZARD WARNING
Soda lime is CORROSIVE and can cause severe burns; also dangerous to eyes and skin. AVOID SKIN CONTACT. AVOID CONTACT WITH WATER. WEAR EYE PROTECTION.

A20 (p59) HAZARD WARNING
Soda lime is CORROSIVE and can cause severe burns; also dangerous to eyes and skin. AVOID SKIN CONTACT. AVOID CONTACT WITH WATER. WEAR EYE PROTECTION.

A22 (p61) HAZARD WARNING
Sodium chlorate(I) solution is corrosive. AVOID SKIN CONTACT. WEAR EYE PROTECTION.

A23 (p62) HAZARD WARNING
Alcohol is highly flammable. KEEP AWAY from naked flame. WEAR EYE PROTECTION.

A24 (p63) HEALTH CHECK
Check with your teacher that you are able to participate in this activity.

A25 (p64) HAZARD WARNING
Alcohol is highly flammable. KEEP AWAY from naked flame. WEAR EYE PROTECTION.

A28 (p67) HAZARD WARNING
Iodine and Benedict's are harmful to skin and eyes. AVOID SKIN CONTACT. WEAR EYE PROTECTION.

A29 (p68) HAZARD WARNING
Ethanol is highly flammable. KEEP AWAY from naked flame. Biuret is harmful to skin and eyes. AVOID SKIN CONTACT. WEAR EYE PROTECTION.

A30 (p69) HAZARD WARNING
Wear EYE PROTECTION when burning foods. Use a safety screen.

A31 (p70) HAZARD WARNING
Amylase is an irritant. AVOID SKIN CONTACT. Iodine and Benedict's are harmful to skin and eyes. AVOID SKIN CONTACT. WEAR EYE PROTECTION.

A34 (p73) HAZARD WARNING
Scalpels are sharp, handle with care.

A39 (p78) HAZARD WARNING
Take care when heating soil. WEAR EYE PROTECTION. Universal indicator is highly flammable and harmful if swallowed. KEEP AWAY FROM NAKED FLAME.

A40 (p79) HAZARD WARNING
WEAR EYE PROTECTION. BEWARE OF SPITTING HOT PARTICLES!

A41 (p80) HAZARD WARNING
Take care when heating soil. WEAR EYE PROTECTION. BEWARE OF SPITTING HOT PARTICLES!

A43 (p84) HAZARD WARNING
Take care when 'throwing' quadrants.

A45 (p86) HAZARD WARNING
Never culture material from a lavatory, sewage-polluted water, animal cage or any human source e.g. finger nail scrapings.

A46 (p87) HAZARD WARNING
Wash hands thoroughly after touching agar. During incubation, plates SHOULD NOT be completely sealed. After incubation, plates must be completely sealed with tape before observing results.

A47 (p88) HAZARD WARNING
During incubation, plates SHOULD NOT be completely sealed. After incubation, plates must be completely sealed with tape before observing results.

A48 (p89) HAZARD WARNING
Do not remove cotton wool plugs at any time during this activity.

ASSESSMENT TASKS

T1 (p110) HAZARD WARNING
Scalpels are sharp, handle with care.

T2 (p112) HAZARD WARNING
Ethanol is highly flammable. KEEP AWAY from naked flame. Iodine is harmful to skin and eyes. AVOID SKIN CONTACT. WEAR EYE PROTECTION.

T3 (p115) HAZARD WARNING
Wear EYE PROTECTION when burning foods.

T4 (p118) HAZARD WARNING
Benedict's is harmful to skin and eyes. AVOID SKIN CONTACT. WEAR EYE PROTECTION.

T5 (p121) HAZARD WARNING
Take care when heating soil. WEAR EYE PROTECTION. BEWARE SPITTING HOT PARTICLES!

PART A

Teacher's Notes

The following Notes offer a comprehensive cross-referencing guide to the Coordinated Science series. They are organised by Students' Book chapter and then by individual 'spread' eg 1.1, 1.2 etc.

A brief summary of the contents of each spread is followed by two boxed sections: *Students'*; *Teacher's Guide.*

Students' sections provide detailed cross-referencing to the other Students' Books in the series and detail where material of a similar or related nature may be found. This is also presented by spread number/title.

Teacher's Guide sections provide listings and locations of Activities (A), Investigations (I), and Assessment Tasks (T) of a similar or related nature occurring anywhere within the series as a whole.

1 Variety of life

1.1 Living things

Topics covered: characteristics of living things – movement, senses, feeding, respiration, excretion, reproduction and growth.

Living things need food: to provide energy and to provide materials for building new tissue. Plants use energy from the Sun to turn simple substances like carbon dioxide and water into food.

Living things respond to a variety of external forces and changes, including light, sound, and gravity.

Students'
Physics
1.10 The pull of gravity
2.1 Chemical energy – stored in plants
2.8 Energy resources – energy from the Sun
3.7 The eye – a light detector
3.14 Sound waves
Chemistry
11.5 Oxygen needed for respiration
12.2 Carbon in living things
13.4 Air and its importance to living things
13.6 Water and its importance to living things

1.2 Living things and their needs

Topics covered: basic requirements for organisms living in the atmosphere and in water; the biosphere.

On Earth, organisms live within the biosphere. This provides them with warmth, water and essential gases and minerals. The Sun is the source of energy for life within the biosphere.

Students'
Physics
1.21 The atmosphere
2.8 Energy from the Sun
2.22 Human beings need warmth
3.1 Light as a form of energy
Chemistry
10.5 Minerals from the Earth's crust
11.5 Oxygen needed for respiration
12.2 Carbon in living things
13.4 Air and its importance to living things
13.6 Water and its importance to living things

1.3 Sorting and naming
1.4 More about sorting and naming
1.5 Groups of living things

Topics covered: dividing living organisms into groups; using a key; major classifications of living organisms.

Teachers may wish to draw comparisons between the principles used to classify living organisms with those used to classify elements.

Elements can be classified by their properties and by their atomic structures. It was a detailed study of the former which led to an understanding of the latter.

Students'
Physics
5.1 Classifying atoms
Chemistry
2.1 Classifying elements
2.3 Classifying atoms
2.4 The Periodic Table
10.1 Metals and non-metals

Teacher's Guide
Chemistry
A8 Looking at elements

1.6   Protoctista and fungi

Topics covered: Protozoa, fungi and algae.

In general, the simplest organisms are also the smallest. Even simple organisms are built from complex compounds using elements extracted from the air.

Students'
Physics 1.1   Relative sizes **Chemistry** 6.1   Living organisms are the products of chemical change 13.3 Materials for living organisms come from the atmosphere

Teacher's Guide
Biology A1   Microscopes and how to use them A2   Making microscope slides A3   Looking at cells A4   Measuring cells A5   Investigating soil life

1.7   Animals without backbones I

Topics covered: invertebrates: coelenterates, flatworms, annelids, worms and molluscs.

Many invertebrates have no rigid materials in their structure. Their form is maintained by liquid pressure. When they feed and move a range of physical principles may be put to use. These include movement by wave action, movement by jet propulsion of water, filter feeding, and suction for collecting food or clinging to surfaces.

Students'
Physics 1.12 The principle of jet propulsion 1.20 Liquid pressure 1.21 Suction and atmospheric pressure 2.7   Pressure transmitted through liquids 3.8   Wave motion **Chemistry** 1.8   Separating mixtures by filtering

1.8   Animals without backbones II

Topics covered: arthropods: crustaceans, insects, arachnids, myriapods.

An arthropod has a tough skin called a cuticle. It has a high carbon content and is strong but flexible. It provides firm anchorage for muscles so rapid movement of wings and limbs can be achieved. Arthropods can have simple eyes with one lens, or compound eyes with thousands of lenses. Some arthropods have stings, which may be acid or alkaline.

Some insects have wings or body parts which are brightly coloured. The colours are seen because only some of the spectral colours in sunlight are reflected by the surface. The others are absorbed.

Students'
Physics 1.16 Strength of materials 2.6   Levers; lever action of muscles 3.2   Surfaces reflect light 3.4   Colours in the spectrum 3.7   The eye 3.12 Seeing colours **Chemistry** 9.4   Insect stings 12.2 Carbon in living things

Teacher's Guide
Biology A37 Pollution **Physics** A9   The Amazing Superflea!

1.9   Animals with backbones I

Topics covered: vertebrates: fish, amphibians and reptiles.

Fish propel themselves forward by pushing a mass of water backwards with their tails. They have streamlined shapes to reduce friction from water. They keep afloat because the water pressure underneath them is greater than the pressure above. This produces a resultant upward force (upthrust) which is just enough to balance the weight of the fish.

Like all living things, fish need a supply of oxygen for respiration. Pond, river and sea water all contain dissolved oxygen. Fish use their gills to extract this oxygen. The oxygen diffuses through a thin membrane into the bloodstream.

Students'
Physics
1.11 Fluid friction
1.12 The principle of jet propulsion
1.18 Liquid pressure can keep things afloat
Chemistry
1.4 Diffusion of gases
1.7 Solubility of oxygen in water
11.5 Oxygen and respiration
13.4 The importance of oxygen to living things

Teacher's Guide
Biology
A21 A fishy problem

1.10 Animals with backbones II

Topics covered: warm-blooded vertebrates: birds, mammals.

Birds and mammals have skeletons of bone – a strong, low density material containing the metal calcium. The skeleton provides firm anchorage for muscles so that rapid movement of limbs or wings can be achieved.

The life-maintaining chemical reactions in birds and mammals require a constant temperature environment. If the temperature is not maintained, the efficiency of the animal as an 'engine' drops. Animals have many mechanisms for keeping body temperature steady; for example, air trapped in fur or feathers for insulation, the evaporation of sweat for cooling.

Birds and mammals use their eyes and ears as transducers to get information from the world around them. In mammals, the outer ear often has a concave surface to reflect, focus and channel sound waves.

Students'
Physics
1.2 Densities of solids
1.18 Strength of materials
2.4 The human engine
2.5 Efficiency of the human engine
2.12 Body temperature
2.19 Air as a poor conductor of heat
2.25 Cooling effect of sweating
3.3 Concave reflectors
3.16 Reflection of sound waves
5.14 Transducers
Chemistry
8.3 Effect of temperature on rate of reaction
10.5 Calcium in the Earth's crust

Teacher's Guide
Biology
A16 Bones
A17 Natural and man-made structures
Physics
A5 Ten times bigger

1.11 Plants

Topics covered: non-flowering and flowering plants.

Plants are built from compounds containing carbon, nitrogen, hydrogen and oxygen. Air provides the starting materials for these compounds. Sunlight provides the energy for the reactions which make them. The energy in sunlight is absorbed by chlorophyll – a substance which makes plants green.

Students'
Physics
2.8 Plants absorb energy from the Sun
3.1 Light as a form of energy
Chemistry
2.1 Compounds
6.1 Chemical change
12.2 Carbon in living things.
13.3 Plants built from elements in the air

2 Cells, heredity, and evolution

2.1   What are cells?

Topics covered: structure of animal cells; structure of plant cells.

The word cell is often used to describe a basic 'unit' of activity. Electric cells are 'units' which can be joined together to make a battery.

Living cells are mainly built from elements which come from the air – carbon, nitrogen, hydrogen and oxygen. They contain hundreds of compounds, some of which are dissolved in water. Water pressure gives cells their form and firmness.

Teachers may wish to draw comparisons between cells as the basic building blocks of living organisms, and atoms as the basic building blocks from which all materials are made. Cells, which are made of atoms, are of the order of 10^8 larger than atoms.

Students'
Physics 1.1   Size of cells 1.20 Liquid pressure 4.2   Electric cells **Chemistry** 1.1   All matter is built from particles (atoms) 1.5   Mixtures; solutions 2.1   Atoms, elements, compounds 12.2 Carbon in living things 13.4 Air provides the materials from which cells are built 13.6 Water in cells

Teacher's Guide
Biology A1   Microscopes A2   Making microscope slides A3   Looking at cells A4   Measuring cells A11 Transport tissues in plants

2.2 Cells, tissues, and organs

Topics covered: cell division; tissues; organs; organ systems; organisms.

Most cells in the body are specialized. For example, nerve cells transmit electrical pulses to monitor and control the body; red blood cells carry oxygen round the body. All cells must be supplied with oxygen so that respiration can take place.

Teachers may wish to draw comparisons between the way specialized cells of the nervous systems combine to build a fully functioning organism and the way transducers and other components are combined to produce an electronic system.

Students'
Physics 4.2   Electrical conductors 5.11 Electronic systems 5.14 Advanced electronic systems **Chemistry** 11.2 Oxygen in cells

Teacher's Guide
Biology A3   Looking at cells A11 Transport tissues in plants

2.3   In and out of cells

Topics covered: diffusion; materials transported in and out of cells by diffusion; osmosis; osmosis in plant cells.

Materials pass in and out of cells by diffusion. It is the process by which cells gain the oxygen they need for respiration and lose the waste carbon dioxide they produce.

In plants, water passes from cell to cell by osmosis. This is a special kind of diffusion in which a membrane acts as a filter, allowing water molecules to pass through but stopping larger molecules. Plant cells are kept firm by the pressure of the water inside them.

2.4 Cell division
2.5 Heredity and variation
2.6 Chromosomes and genes
2.7 More about chromosomes

Topics covered: chromosomes; cell division; sex cells and fertilization; inherited and acquired characteristics; continuous and discontinuous variations; DNA and genes; dominant and recessive genes; dominant and recessive characteristics.

Cells are complex chemical factories in which new compounds are formed from the simple substances brought in. The 'building instructions' for each cell are stored in coded form by a compound called DNA.

Discontinuous variation is a feature of atomic structure. The atoms of each element show variation in their different isotopic forms.

2.8 Patterns of inheritance I
2.9 Patterns of inheritance II

Topics covered: alleles; phenotypes and genotypes; the effect of the environment on phenotypes; mutations; monohybrid crosses; codominance; sex-linked inheritance.

The phenotypes of living organisms are controlled partly by their genotypes and partly by environmental factors. For example, two plants may have the same genotype, but if one is deprived of nitrogen, its growth characteristics will be altered.

Some environmental factors cause mutations. Nuclear radiation is one example. The ionization causes in living cells can alter their genetic code.

Certain types of colour blindness are sex-linked. The affected individual is unable to distinguish between some parts of the visible spectrum which a 'normal' eye could separate.

<table>
<tr><td colspan="1">Teacher's Guide</td></tr>
</table>

Biology
A6   A model for genetics

Physics
A33 Radioactivity

2.10   DNA and the genetic code
2.11   Genetic engineering

Topics covered: structure of DNA; DNA and cell division; DNA codes; manipulating DNA; uses of genetic engineering.

Amino acids are 'built' from the simple substances carbon, hydrogen, oxygen and nitrogen. Different amino acids are linked to give protein molecules. It is the sequence of bases along a DNA molecule which tells a cell which amino acids to link to form protein.

Manipulation of genes allows scientists to form organisms which have special characteristics. One application of genetic engineering is the production of useful substances including fuels and plastics.

Some applications of genetic engineering may raise questions of morality.

<table>
<tr><td>Students'</td></tr>
</table>

Chemistry
11.3 Fertilizers
13.4 Nitrogen in living things

<table>
<tr><td>Teacher's Guide</td></tr>
</table>

Chemistry
A26 Testing foods for nitrogen

2.12   Biotechnology

Topics covered: biotechnology: uses and products.

Living organisms, particularly micro-organisms, may be used to produce useful substances. These include alternatives to fossil fuels.

Biotechnology is used in the disposal of sewage and the management of our water resources.

<table>
<tr><td>Students'</td></tr>
</table>

Physics
2.8   Energy resources

Chemistry
12.8 Alcohol by fermentation
13.7 Our water supply

<table>
<tr><td>Teacher's Guide</td></tr>
</table>

Biology
A7   An alcohol problem
A44 Microbiology: safety
A45 Growing and studying bacteria
I7   Investigating artificial meat
I10 Investigating fermentation

2.13   Evolution I
2.14   Evolution II

Topics covered: evolution; natural selection; artificial selection; the fossil record; extinction and the formation of new species.

Environmental factors influence the course of evolution. Many organisms have evolved surface markings which reflect light in a pattern similar to their background – or in marked contrast to it.

Evolution is a slow process. The great age of our Earth has allowed a wide variety of organisms to evolve.

Sedimentary rocks were formed by the action of pressure on sedimentary deposits from seas and rivers. Many fossils are found in such rocks. Understanding the structure of the Earth enables us to estimate the age of fossils.

<table>
<tr><td>Students'</td></tr>
</table>

Physics
3.12 Seeing colours
6.6   Age of the Solar System

Chemistry
13.2   Records of the Earth's past
13.11 Sedimentary rocks
13.13 Interpreting sedimentary layers

<table>
<tr><td>Teacher's Guide</td></tr>
</table>

Biology
A5   Looking at variation
A6   A model for genetics

3 Flowering plants

3.1 Flowering plants
3.2 More about plant nutrition

Topics covered: the parts of a plant; photosynthesis; how plants use glucose; plants and minerals.

A plant supports a widely-spread weight of foliage on a narrow stem or trunk. This structure is inherently unstable, so firm roots are needed to resist turning effects produced by wind or an uneven weight distribution.

During photosynthesis, plants take in simple substances from the air and use them to make food. Some is used to build new plant tissue. Some is used during respiration to give the plant energy. Some is stored.

Energy stored in plants originally came from the Sun. Fossil fuels store energy originally absorbed by living plants.

Students'
Physics 1.16 Turning effects of forces 1.17 Centre of gravity and stability 2.1 Types of energy 2.2 Energy changes 2.8 Plants store energy from the Sun 3.1 Light is a form of energy **Chemistry** 11.5 Plants use oxygen during respiration 12.2 Carbon in plants 12.3 Fossil fuels made from crushed plants 13.4 Plants build from substances in air 13.6 Plants need water for photosynthesis

Teacher's Guide
Biology A3 Looking at cells A8 Photosynthesis and oxygen A9 Chlorophyll and photosynthesis A10 Light, carbon dioxide and photosynthesis I1 Light and photosynthesis T2 Investigating photosynthesis in leaves

3.3 Leaves

Topics covered: structure of a leaf; leaves suited to photosynthesis.

Leaves are green because of the chlorophyll they contain. Chlorophyll is the catalyst for photosynthesis reactions in which the Sun's energy is used to turn simple substances into food. Leaves take in carbon dioxide through their leaves, and give out oxygen and water.

Leaves have large surface areas to absorb as much solar energy as possible. However, a large surface area means a potentially large evaporation rate so many leaves have waxy coatings to reduce water loss.

Leaves can be damaged, and whole areas of forest destroyed, by acid rain.

Students'
Physics 2.24 Factors affecting evaporation 3.1 Light is a form of energy 4.23 Acid rain **Chemistry** 11.7 Acid rain 12.2 Chlorophyll; photosynthesis; gas exchange in leaves

Teacher's Guide
Biology A13 Measuring transpiration A14 Stomata A37 Pollution (dirty leaves) A38 More pollution (acid rain) I2 Leaf colour and photosynthesis I3 Transpiration through leaves T2 Investigating photosynthesis in leaves

3.4 Transport and support in plants
3.5 Transpiration

Topics covered: transport system in roots and stems; support in plants; transpiration.

In a plant, the transpiration stream flows upwards from roots to leaves. This keeps all living cells supplied with water and minerals.

In hot weather leaves are kept cool by the evaporation of water from their surfaces. In large areas of forest, the transpiration stream can have a significant effect on the local climate.

Plants take in water and minerals through their roots by diffusion. The uptake of minerals by plants can be tracked using radioactive tracers. Radioactive minerals from contaminated soil can become incorporated into the structure of plants, and make them a hazard to humans unless the half-life is very short.

Food and other substances are carried round plants dissolved in water. The soft parts of plants are kept firm or turgid by the pressure of the water in their cells. The harder parts form into wood. Wood is a low density material with great strength and flexibility.

Students'
Physics
1.18 Strength of materials
1.20 Water pressure
2.11 Diffusion
5.2 Radioactive contamination of soil
5.3 Dumping nuclear waste
5.5 Half-life; decay hazards
Chemistry
1.1 Diffusion
1.5 Water as a solvent for sugar
2.9 Radioactive tracers
13.6 Water in plants

Teacher's Guide
Biology
A12 Osmosis
A11 Transport tissue in plants
A13 Measuring transpiration
A14 Stomata
A39 & A40 Comparing soil samples
I3 Transpiration through leaves
I7 Investigating garden soil
T5 Measuring the water, humus and mineral content of soil
Chemistry
A30 How acid rain kills

3.6 Flowers
3.7 How fruits and seeds begin
3.8 More about fruits and seeds

Topics covered: parts of a flower; types of flower; pollination by wind and insects; fertilization; dispersal by wind, animals.

Insects transfer pollen as they move from flower to flower, attracted by the colours, in daylight reflected by the flower or by the scents which diffuse through the air. Some pollen is spread by wind. The tiny pollen particles have very low terminal speed so they remain airborne for a long time.

The seeds from some plants are spread by wind. Dandelion and thistle seeds offer very high air resistance for their weight, so their terminal speeds are low. Wind will move them long distances before they reach the ground.

Students'
Physics
1.11 Air resistance; terminal speed
2.11 Scents spread by diffusion
3.4 Colours in white light (daylight)
3.12 Seeing colours
Chemistry
1.1 Diffusion
1.4 Diffusion of gases

Teacher's Guide
Physics
I3 Investigating parachute design

3.9 How seeds grow into plants

Topics covered: germination and the development of young plants.

To germinate, seeds need air, water and warmth. The reactions involved in germination require a temperature within certain limits – normally 5 °C to 45 °C.

Students'
Physics
2.12 Temperature
Chemistry
6.3 The effect of temperature on a reaction
13.4 The importance of air to plants
13.6 The importance of water to plants

Teacher's Guide	*Students'*
Biology A15 Germination and growth	**Physics** 1.10 The pull of gravity 3.1 Light as a form of energy **Chemistry** 12.2 Plants need light for photosynthesis 13.6 The importance of water to plants

3.10   Plant senses

Topics covered: the response of plants to light, gravity and water.

Plants respond to the Earth's gravitational field, so that roots grow down and shoots grow up. They also grow so as to maximize the plant's intake of light energy and water.

4 Support, respiration, and breathing

4.1   Skeletons

Topics covered: liquid 'skeletons'; external skeletons; internal skeletons; bone; the human skeleton.

Animals without skeletons keep their shape and form because of the pressure of the liquid inside them. An exoskeleton provides firm anchorage of muscles, so large movements of limbs can be achieved. Some exoskeletons have evolved to reflect certain colours only.

Internal skeletons are made of bone – a strong, low-density material containing the metal calcium. They also provide firm anchorage for muscles so that limbs can act as machines.

Students'
Physics
1.2   Density of materials
1.18 Strength of materials
1.20 Pressure in liquids
2.6   Limbs as machines
3.1   Surfaces which reflect light
3.4   Colours in white light (daylight)
3.12 Seeing colours
Chemistry
13.8 Hard water: calcium for bones and teeth

Teacher's Guide
Biology
A16 Bones
A17 Natural and man-made structures

4.2   More about the human skeleton

Topics covered: the human skeleton; the backbone; types of joint. At movable joints, lubricating fluid is trapped between the bones to reduce friction.

In the backbone, the vertebrae are separated by discs of cartilage. These are elastic and reduce the effect of vibrations and acceleration forces.

Even so, there is a limit to the acceleration forces that the human frame can withstand. Fairground rides stay within it – at around '3g'.

Students'
Physics
1.8   Fairground acceleration forces
1.11 Friction between surfaces
1.18 Elasticity; strength of materials
3.15 Vibrations
Biology
A16 Bones

4.3   Muscles and movement

Topics covered: voluntary and involuntary muscles; cardiac muscles; pairs of muscles in limbs.

Muscles change stored chemical energy into kinetic energy. When chemical changes take place inside a muscle, the result is a physical change: the muscle contracts.

Human limbs are machines which act as movement magnifiers. A large input force gives a small output force, but greatly increased movement.

Muscles are amplifiers. A low energy input signal from the nervous system produces a relatively high energy output when the muscle moves.

Students'
Physics
1.9   Force
1.16 Turning effect of a force; moments
2.1   Chemical energy
2.2   Energy changes
2.4   Muscles are human 'engines'
2.6   Limbs as machines
5.11 Amplifiers
Chemistry
6.1   Physical and chemical change

Biology
A18 Muscle fatigue
A24 Pulse and breathing rates
I4   Investigating muscle fatigue and
      recovery

Physics
A10 Measuring your power output

4.4 Metabolism, enzymes, and respiration

Topics covered: metabolism; enzymes; respiration and breathing; the human respiratory system.

Complex organisms are kept alive through a huge number of chemical reactions. Some of these break complex molecules into simpler substances: others synthesize complex molecules from simpler ones.

Enzymes are 'biological catalysts'. Their action can be compared with the catalysts used in industrial processes.

Enzymes speed up reaction rates. Many require particular conditions of acidity or alkalinity. Many require a body temperature close to 37 $^\circ$C. Life may be threatened if this temperature is not maintained.

Physics
2.4   The human engine
2.5   Power and efficiency of the human
      engine
2.12 Temperature
2.13 Measuring body temperature
2.22 Risks from hypothermia
2.25 Temperature control in humans

Chemistry
6.4   Combination reactions
6.4   Enzyme action in ferment
8.1   Rates of reactions
8.3   Factors affecting the rate of a reaction
8.4   Factors affecting a rate of enzymes
8.7   More about catalysts
9.1   Acids and alkalis
11.5 Energy from respiration

Biology
A25 The functions of perspiration
A26 Keep it cool
A31 Enzymes and digestion
I12 The browning of apples and pH

Chemistry
A15 Catalysts at work
T3   Catalysing a reaction

4.5 Aerobic and anaerobic respiration

Topics covered: aerobic and anaerobic respiration; oxygen debt.

During respiration chemical energy in food is changed into other forms, such as heat and kinetic energy.

Respiration in muscles is normally aerobic. It produces carbon dioxide and water. During vigorous exercise, when more oxygen is needed, anaerobic respiration may occur. This produces lactic acid.

The action of ATP as an 'energy store' can be compared with physical counterparts including electric cells and capacitors.

Physics
2.2   Energy changes during a weight lift
2.4   The human engine
2.5   Efficiency of the human engine

Chemistry
6.4   Oxygen for energy
9.1   Acids
11.5 The importance of oxygen; respiration in
      humans

Biology
A19 Demonstrating respiration
A20 Measuring respiration
A22 Anaerobic respiration
I5   Investigating respiration rates

4.6 Lungs and gas exchange

Topics covered: the lungs; gas exchange; breathing action; gas content in the lungs.

In the lungs, oxygen passes into the blood by diffusion, and carbon dioxide passes out. Air is forced into the lungs by atmospheric pressure when an increase in volume of the chest cavity causes a reduction of pressure inside.

When gases exchange in the lungs, they dissolve in liquid before moving across the lung lining by diffusion. For a high rate of diffusion, lungs have large inner surface areas. Evaporation and rate of reaction are two other processes which depend on surface area.

Gas exchange in the lungs lowers the oxygen content of the air and raises the water vapour and carbon dioxide content. However, condensation and the effects of photosynthesis in plants keep the balance of gases in the atmosphere approximately constant.

A build-up of water vapour in a room can make conditions uncomfortable for the occupants. Air pollution, and lack of regular air changes, can seriously upset the balance of gases taken into the lungs. The results can be damaging to health, or even fatal.

Students'

Physics
1.21 Using atmospheric pressure
2.10 Need for air changes in a room
2.11 Diffusion
2.18 Pressure/volume relationship for a gas
2.25 Humidity

Chemistry
1. 1 Diffusion
1.4 Pressure/volume relationship for a gas; diffusion
12.1 Testing for carbon dioxide using limewater
12.2 Carbon dioxide in balance – the carbon cycle
13.4 Gases in the air
13.5 Firemen and astronauts use oxygen from cylinders
13.6 Water vapour in balance – the water cycle

Teacher's Guide

Biology
A23 Lung volume
A24 Pulse and breathing rates
I6   Fitness and lung volume

5 Circulation, temperature control, and excretion

5.1 The heart
5.2 What makes blood flow
5.3 Heart disease

Topics covered: structure and action of the heart; arteries and veins; the circulatory system; heart disease.

The heart pumps blood at high pressure through a system of arteries and veins. These are made from elastic and non-elastic fibres. One function of the circulatory system is to transport the oxygen needed for respiration in cells and the carbon dioxide produced as a result.

Teachers may wish to draw comparisons between the circulatory system and an electrical circuit:

- the heart as the 'prime mover'
- the pressure drop round the system
- the resistance offered by arteries and veins
- the parallel branches of the system
- the one-way 'diode' action of the valves in the heart and veins.

Students'
Physics
1.18 Elastic materials
1.20 Liquid pressure
2.7 Transmitted pressure; hydraulic machines
4.5 Resistance
4.6 Dangers from electric shocks
4.7 Parallel circuits
5.9 Diodes
Chemistry
11.5 Oxygen and carbon dioxide carried by blood

Teacher's Guide
Biology
A24 Pulse and breathing rates
Chemistry
I10 Butter or not

5.4 What blood is

Topics covered: plasma; red and white cells; platelets; red cells carry oxygen.

There are about five litres of blood in the body. Blood is a mixture of substances: mainly water, but with other materials in suspension or dissolved in it. Oxygen is carried in the blood by haemoglobin in red blood cells. Carbon monoxide will also combine with haemoglobin and can starve the blood of oxygen.

Students'
Physics
1.1 Volume; the litre
Chemistry
1.5 Mixtures; solutions; suspensions
11.5 The importance of oxygen to living things
12.4 The effect of carbon monoxide on haemoglobin
13.6 Water in blood

5.5 What blood does
5.6 How oxygen and food reach cells

Topics covered: materials and heat carried by blood; fighting disease; tissue fluid; lymph.

Blood distributes heat round the body. It also carries oxygen to the body's cells and takes away the carbon dioxide produced by respiration. Food and oxygen pass from blood capillaries into cells by diffusion. Waste materials pass from cells into the blood in the same way.

Students'
Physics
2.11 Diffusion
2.21 Heat capacity; using water to store and transport heat
Chemistry
1.1 Diffusion
1.5 Mixtures; solutions; suspensions
11.5 The importance of oxygen to living things
12.2 Respiration produces carbon dioxide
13.6 Water in blood

5.7 Homeostasis

Topics covered: tissue fluid as an internal environment; homeostasis; feedback; the organs of homeostasis; osmoregulation; kidney dialysis machines.

The body has a variety of complex mechanisms for maintaining steady conditions within it. Some of these are chemical, such as the control of glucose levels in the blood. Some are physical, such as the cooling effect produced by sweating. Sometimes the body is unable to maintain steady internal conditions by itself. For example, a large loss of body heat may produce hypothermia. Malfunctioning kidneys may need to be aided by a dialysis machine. This effectively filters blood by allowing waste materials to diffuse out.

The body's homeostasis mechanisms rely on feedback. Sensors, either physical or chemical, detect change and affect the control processes accordingly. Teachers may wish to compare the body's feedback mechanisms with those used in thermostats and electronic systems such as amplifiers.

Students'
Physics
2.15 Thermostats
2.16 Thermostatic radiator valve
2.22 Coping with cold; risks of hypothermia
2.25 Cooling effect of evaporation; sweating
5.14 Transducers and electronic systems
Chemistry
1.1 Diffusion
1.8 Separating mixtures by filtering
5.3 Concentration of a solution

Teacher's Guide
Biology
A27 Sensitivity to temperature

5.8 Skin and temperature

Topics covered: the functions and structure of skin; skin colour; keeping a steady body temperature, sweating and shivering; body hair.

The body must maintain a core temperature close to 37 °C to survive. In hot conditions, sweating increases the loss of heat from the body, but high humidity reduces the effect. In cold conditions, loss of heat can be reduced by extra insulation, and by exercise. Like all engines, the body produces waste heat when it is working.

Students'
Physics
2.4 The human engine
2.5 Human (and other) engines produce waste heat
2.13 Measuring body temperature
2.19 Using trapped air as an insulator
2.22 Reducing loss of body heat in extreme conditions
2.25 The cooling effects of evaporation; humidity
3.11 Reducing heat losses by radiation
Chemistry
12.2 Energy released by respiration
13.6 Water lost from the body must be replaced

Teacher's Guide
Biology
A25 The functions of perspiration
A26 Keep it cool
A27 Sensitivity to temperature
Physics
A15 Testing the insulation of a coffee cup
A21 Polar bears and energy conversion
I7 Two pairs or one?
T3 Comparing cooling rates

5.9 Excretion

Topics covered: excretory organs; nephrons.

The body produces waste substances which are carried in the blood. The lungs remove carbon dioxide from the blood. The kidneys remove urea and water by filtering the blood. After filtering, the kidneys return some water and other useful substances to the blood. Without this reabsorption, the body would become dehydrated in a very short time.

Students'
Chemistry
1.5 Mixtures
1.8 Filtering
12.2 Waste carbon dioxide is produced by respiration
13.6 Water losses from the body must be replaced

6 Feeding and digestion

6.1   Food
6.2   More about food

Topics covered: why food is needed; types of food and their main constituents; vitamins and minerals; energy requirements; a balanced diet.

Food is a mixture of substances. The body turns these into the compounds which form new cells and provide energy . The body needs minerals and vitamins. Some become incorporated into the body's structure; others are needed to speed up chemical reactions in the body.

The energy values of different foods can be measured by burning them and finding their heating effect on water.

The body is an engine, with food as its fuel. If the body's energy input exceeds its energy output, the surplus 'fuel' is turned into fat.

Students'
Physics
2.1   Work and energy; joules; food as an energy source
2.2   The body changes chemical energy into heat and kinetic energy
2.4   The human body as an engine; food as fuel
2.5   The efficiency and power of the human body
Chemistry
1.5   Mixtures
2.1   Compounds
9.2   Acids in food
10.5 Calcium and iron
11.3 Fertilizers and food production
11.9 Sodium chloride in food
12.1 Carbon compounds in proteins and carbohydrates
12.2 Glucose as the body's fuel

Teacher's Guide
Biology
A28 & A29 Food tests
A30 Measuring the energy value of foods
T3   Measuring the energy content of dried bananas
T4   Testing for glucose
Chemistry
A26 Testing foods for nitrogen

6.3   Diet health I
6.4   Diet health II

Topics covered: the importance of a balanced diet; energy requirements; obesity and anorexia; risks – overeating, salt, sugar, fat; need for fibre.

As an engine, the body is not very efficient. Up to 85% of the energy it releases is wasted as heat. Carbohydrates and fats are the body's main 'fuels'. Surplus 'body' fuel is turned into fat. Excess fat, salt, or sugar can cause health problems.

Proteins are carbon compounds mainly used for growth. The body uses them as an energy source.

Students'
Physics
2.1   Work and energy; joules; food as energy source
2.2   The body changes chemical energy to heat and kinetic energy
2.4   The human body as an engine; food as fuel
2.5   The efficiency and power of the human body
2.21 Specific heat capacity of water
Chemistry
11.1 Nitrogen in proteins
11.9 Sodium chloride in the diet
12.1 Carbohydrates and proteins
12.2 Glucose as the body's fuel

Teacher's Guide
Biology
A28 & A29 Food tests
A30 Measuring the energy value of foods
T4   Testing for glucose
Chemistry
A26 Testing foods for nitrogen

6.5   Food additives

Topics covered: colourings, flavourings, preservatives and their effects.

Chemicals are frequently added to processed foods to 'improve' the flavour or appearance, or to extend their storage life. Carbon dioxide is the additive which gives drinks their 'fizz'.

Alternative methods of food preservation include freezing and irradiation (controversial).

Students'
Physics
5.3 Food irradiation
Chemistry
11.7 Sulphur dioxide as a preservative
11.9 Sodium chloride as a food additive
12.1 Carbon dioxide in drinks
13.5 Liquid nitrogen used for freezing food

Teacher's Guide
Biology
A32 Investigating the sense of taste
Chemistry
I1 Colourful food
I5 Fizzy facts

6.6   Teeth

Topics covered: structure of a tooth; decay, gum diseases and how to prevent them.

Teeth are made partly from hard, inelastic materials containing the metal calcium. They are held in their sockets by elastic fibres. Tooth enamel dissolves in the acids produced by eating sugary foods.

Students'
Physics
1.18 Strength of materials; elastic and inelastic materials
Chemistry
1.5 Dissolving
9.2 The effects of acids
10.5 Calcium
13.7 Fluoride in the water supply
13.8 Hard water

Teacher's Guide
Chemistry
I12 Investigating the hardness of water

6.7   Digestion
6.8   A closer look at digestion

Topics covered: stages of digestion; digestive enzymes; digestion in the mouth; swallowing; digestion in the stomach and intestine.

In the digestive system, food is chemically changed by enzymes to simpler chemicals dissolved in water. Digested and undigested foods are separated in the small intestine. Digested food passes into the blood. Water is removed from undigested food in the colon. Food is moved through the alimentary canal by peristalsis – a 'transverse wave' muscle action. Fermentation is one example of enzyme action.

Students'
Physics
3.8 Wave action
Chemistry
1.5 Dissolving; solutions
1.8 Separating mixtures
6.1 Chemical change
6.4 Enzyme action in fermentation
8.7 Liver contains the catalyst, catalase
9.4 The stomach contains hydrochloric acid

Teacher's Guide
Biology
A7 An alcohol problem
A31 Enzymes and digestion
I10 Investigating fermentation
Chemistry
I2 Pain and pH

6.9   Absorption and the liver

Topics covered: food absorption in the small intestine; villi; the liver as a chemical factory and energy store.

The small intestine has a large surface area. Digested food passes by diffusion though this surface into the blood. The food is carried to the liver for processing. The liver is a chemical factory, forming new compounds from the substances being brought in. It also acts as a heat store.

Students'
Physics
2.11 Diffusion
2.21 Storing heat – heat capacity
Chemistry
1.1 Diffusion
2.1 Compounds
6.1 Chemical change
12.2 Body tissues are built from carbon compounds

7 Senses and co-ordination

7.1 Touch, taste, and smell

Topics covered: the sense organs; the skin; the tongue; the nose.

The body has many detectors for sensing conditions in the outside world. Detectors in the skin react to pressure and temperature. The ears respond to sound waves. The eyes respond to light. The nose lining detects chemicals which have diffused through the air. Taste buds detect chemicals dissolved in saliva.

There are similarities between many electronic systems and the body's sensor system. Transducers convert energy inputs into electrical signals which are processed within the system.

Students'
Physics
1.20 Pressure
2.11 Diffusion
2.13 Temperature sensors; thermometers
3.14 Detecting sound waves
3.17 Distinguishing pitch
5.12 Electronic circuits which detect heat and light
5.14 Sensors and systems
Chemistry
1.1 Diffusion
1.4 Diffusion of gases
1.5 Dissolving

Teacher's Guide
Biology
A27 Sensitivity to temperature
A32 Investigating the sense of taste

7.2 The eye
7.3 Vision

Topics covered: how the eye focuses light; protecting the eye; structure of the eye; stereoscopic vision; light-sensitive cells in the retina.

In a human eye, a convex lens system focuses an image on the retina. Cells in the retina contain light-sensitive chemicals. There are similarities between the eye and a camera. In the eye, however, focusing adjustments are made by altering the thickness of an elastic lens.

Students'
Physics
1.18 Elastic materials
3.4 Bending light
3.6 Convex lenses
3.7 The human eye; the camera
Chemistry
6.4 Light-sensitive chemicals
4.3 Contact lenses are made of plastic

Teacher's Guide
Biology
A34 Eyes and vision I
A35 Eyes and vision II
Physics
A19 Tracing rays through a glass prism
A22 Measuring the focal length of a convex lens
I9 Object-image distance in lenses
T4 The image formed by a convex lens

7.4 Ears, hearing and balance

Topics covered: sound waves; structure of the ear; how the ear detects sounds; semicircular canals and balance.

Sounds are rapid pressure changes. They travel through the air (and other media) as compression waves. They are detected by the inner ear which sends electrical signals to the brain. The system can distinguish between incoming waves of different frequencies and amplitudes.

There are similarities between many electronic systems and the body's sound detecting system. Transducers convert energy inputs into electrical signals which are then processed.

The inner ears also contain the organs of balance.

7.5   More about senses

Topics covered: hunger and thirst; muscles and joints; sensing temperature differences; loudness of sound; clarity of vision; the blind spot.

Water and food are essential for life. Receptors in the brain receive signals from sensors in the body. The sensations of thirst and hunger 'encourage' us to take in more food and water.

Thermometers measure temperature. Sensors in the skin help us to judge temperature differences.

In some circumstances, sound is a form of pollution. High energy sound waves may cause damage to the ear.

7.6   The nervous system
7.7   More about reflexes
7.8   The brain

Topics covered: the nervous system; nerve cells and their action; reflex and voluntary actions; the brain. The brain and nervous system are like a complex electronic system – processing input signals from a variety of sensory nerves and producing outputs as a result. Nerve fibres conduct the signals. Teachers may wish to draw comparisons with optical fibres which transmit signals in the form of light pulses.

If absorbed bodily, some substances damage the nervous system. Lead and mercury are examples.

7.9   Hormones

Topics covered: endocrine glands; hormones; the functions of hormones.

Hormones are chemicals which control a variety of chemical reactions inside the body.

8 Reproduction

8.1   Two kinds of reproduction

Topics covered: asexual and sexual reproduction; external and internal fertilization.

The ability to reproduce is a principal feature of living organisms. So far it is an ability which cannot be matched by even the most advanced electronic systems.

Students'
Physics
5.16 Machines that think

8.2   Human reproduction I
8.3   Human reproduction II
8.4   Life before birth
8.5   Birth, and birth control

Topics covered: sexual development; sex organs; sexual intercourse; ovulation, fertilization and the menstrual cycle; the development of an embryo; from foetus to birth; health of mother and baby; birth control.

Llke all living things, a developing embryo must respire to survive. Food, oxygen and other materials are passed to the embryo through the umbilical cord. They pass from the mother's blood to the embryo's by diffusion. Some are chemically changed into new body tissue. If pollutants are present, they pose a special hazard to the embryo at this early stage.

In the womb, a developing baby is protected from vibration by the cushioning effects of water.

Students'
Physics
2.11 Diffusion
3.15 The damaging effects of vibrations
5.2 Dangers of nuclear radiation
5.3 Dangers from radioactive waste
5.5 Dangers of radioactive strontium and iodine
Chemistry
1.1 Diffusion
6.1 Chemical change
11.5 Respiration requires oxygen and produces carbon dioxide
12.2 Carbon compounds form new body tissue
12.8 The effects of alcohol on an unborn baby
13.4 Pollutants in air
13.6 Water in the human body
13.7 Pollutants in water

8.6   Embryos, babies and children

Topics covered: test-tube babies; embryo testing; growth and development; puberty.

Modern techniques allow human ova to be fertilized *in vitro* and then transplanted into the natural or surrogate mother. This raises moral issues.

Chemical testing of amniotic fluid and non-invasive techniques may reveal defects in the developing embryo. The question as to whether or not embryos should be aborted under these circumstances arouses strong arguments.

Students'
Physics
3.18 Ultrasound and its uses

8.7   Human populations
8.8   Feeding the world's billions

Topics covered: the increase in world population; problems of population growth; why millions starve.

Humans get their energy from food. It takes more than twenty times the land area to provide this energy in the form of meat than as cereals. Yet in countries where food shortages are greatest, there is pressure to rear cattle rather than grow crops because of the better returns in export markets. Developing countries need to earn foreign currency to buy oil. The development of local energy sources offers a partial solution to the problem.

Despite increased crop yields through the use of fertilizers, much of the world still starves. In some areas, deforestation has upset the water cycle, reduced rainfall, and turned agricultural land into desert. If global warming is taking place, the situation could deteriorate.

The need for increased food output can lead to the excessive use of pesticides and fertilizers. These find their way into rivers and are a source of pollution.

Students'
Physics
2.1 Energy in food; the joule
2.8 Alternative energy sources
4.23 Hydroelectric power in the developing world
Chemistry
11.3 Fertilizers
13.6 The water cycle

Teacher's Guide
Biology
A43 Ecology II (populations)
Chemistry
A27 Making a fertilizer
A33 Comparing fuels

8.9   Wild populations

Topics covered: growth of wild populations; factors which limit population growth; predator-prey relationships.

Changes in weather conditions can affect the size of wild populations. Long term changes, such as global warming, may upset the balance between competing species.

Pollution can affect the birth-rate of a species to fall and/or the death-rate to rise. Chemical pollution is particularly serious in a closed environment such as a pond or lake.

Students'
Physics
2.26 Changing the weather
4.23 Acid rain
Chemistry
11.3 Fertilizers
11.7 Acid rain
12.4 Fossil fuels and the environment

Teacher's Guide
Biology
A38 More pollution (acid rain)
Chemistry
A30 How acid rain kills
A31 Living in a greenhouse
A36 Ozone – a saver of life

9 Living things and their environment

9.1 Depending on each other
9.2 More about food chains

Topics covered: producers and consumers; food chains, pyramids and webs; decomposers; trophic levels; flow of energy along food chains; biomass.

Plants take in simple materials such as carbon dioxide and water and chemically change them into more complex compounds. These pass along food chains and are the raw materials from which all animal tissue is built.

All organisms waste energy so only a proportion is passed on to the next organism in the food chain. This limits the length of the chain.

The term 'biomass' is used to describe one type of energy resource. It also has a more precise biological meaning. Teachers may wish to contrast the two.

Students'
Physics
2.1 Energy; the joule
2.5 Humans (and other organisms) waste energy as heat
2.8 Plants and animals obtain energy from the Sun; energy from biomass

9.3 The carbon cycle

Topics covered: photosynthesis, respiration, combustion and decay in the carbon cycle.

Carbon dioxide is taken from the atmosphere by photosynthesis. It is returned to the atmosphere by respiration and combustion. The quantities of oxygen and carbon dioxide in the air stay in balance, but the balance is being disturbed by the excessive use of fossil fuels. This may affect the world's climate.

Students'
Physics
2.4 Exhaust gases from engines contain carbon dioxide
2.26 Carbon dioxide and the greenhouse effect
4.22 Waste gases from thermal power stations contain carbon dioxide
Chemistry
6.4 Synthesis; combustion
11.5 Respiration
12.2 Carbon compounds in living things
12.3 Carbon in living things; photosynthesis; respiration; the carbon cycle
12.4 Fossil fuels produce carbon dioxide when they burn

Teacher's Guide
Biology
A10 Light, carbon dioxide and photosynthesis
A19 Demonstrating respiration
A41 Investigate the functions of soil microbes
Chemistry
A29 Investigating carbon dioxide
A31 Living in a greenhouse?

9.4 The nitrogen cycle

Topics covered: the nitrogen cycle; nitrates formed from nitrogen.

All living things need nitrogen to make proteins. Atmospheric nitrogen enters soil in the form of nitrates. It is returned to the atmosphere by the action of denitrifying bacteria on dead organisms and animal waste.

Students'
Chemistry
11.1 The importance of nitrogen to living things; the nitrogen cycle

---

Teacher's Guide

Biology
A29 Food tests II (proteins)

Chemistry
A26 Testing foods for nitrogen

9.5 The water cycle

Topics covered: evaporation and precipitation of water in the water cycle.

Surface water enters the atmosphere by evaporation. It may exist in the atmosphere as vapour or as clouds (liquid droplets or ice particles). It returns to the ground in the form of rain or snow.

The amount of water vapour in the atmosphere can vary considerably. It is a major feature of weather systems. It also affects the ability of the body to use sweating as a cooling mechanism.

Water for the mains supply is rain-water from reservoirs, rivers and wells. When rain-water percolates through the ground, it dissolves materials in the rock. Some of these make water hard. Hard water causes scaling in kettles, pipes and boilers.

Students'

Physics
2.24 Evaporation
2.25 Cooling by sweating; vapour in the air; humidity
2.26 Human activity and climate

Chemistry
13.4 Gases in the air
13.6 Water; the water cycle; water in living things
13.7 The water supply
13.8 Soft and hard water

Teacher's Guide

Biology
A25 Functions of perspiration
A26 Keep it cool

Chemistry
A7 Testing for water
I12 Investigating the hardness of water

9.6 Planet Earth in danger

Topics covered: the biosphere; threats to the biosphere.

It is now clear that human activity causes great damage to the biosphere. Already many species have become extinct. Without greater understanding and better management, the situation will continue to deteriorate.

Students'

Physics
2.4 Exhaust gases from engines contain carbon dioxide
2.8 Pollution and environmental problems of different energy sources
2.26 Carbon dioxide and the greenhouse effect
4.22 Waste gases from thermal power stations contain carbon dioxide

Chemistry
11.3 Problems with fertilizers
11.7 Sulphur dioxide as a pollutant; acid rain
12.4 Pollution from burning fossil fuels
13.4 Air pollution
13.7 Water pollution
13.9 Soil erosion – the weathering of rocks
13.10 Soil and sediment transport

Teacher's Guide

Biology
A37 Pollution
A38 More pollution

Chemistry
A30 How acid rain kills
A36 Ozone – a saver of life

9.7 Water pollution

Topics covered: causes and effects of water pollution; pollution of food chains.

Water can be polluted by untreated sewage, chemical waste, oil, and the excessive use of fertilizers, pesticides and detergents. Water pollution can reduce the oxygen content of water and threaten pond and river life. It presents a risk to human health because contaminants may enter food chains through fish or plants.

Other forms of pollution include air pollution and nuclear radiation. A principal problem when storing radioactive waste is ensuring that radioactive contaminants do not enter the water supply.

Students'
Physics
5.2 Dangers of nuclear radiation
5.3 Dangers from radioactive waste
5.5 Decay hazards
Chemistry
2.9 Radioactive pollution
6.1 Chemical change
11.2 Ammonia and nitric acid in industry
11.3 Problems with fertilizers
12.2 Making carbon compounds by photosynthesis
13.4 Air pollution
13.7 Water pollution

Teacher's Guide
Biology
A37 Pollution
A38 More about pollution

9.8   More about pollution

Topics covered: air pollution; radiation; litter; noise; reducing pollution; depletion of the ozone layer.

Most air pollution is caused by the burning of fossil fuels. The problem can be reduced by using alternative sources of energy. However, some of these sources, though less polluting, have a major impact on the environment.

Life and health can be threatened by pollutants in water as well as air. Other forms of pollution include radiation, litter and noise.

Pollution is damaging the ozone layer and allowing levels of ultraviolet radiation to rise.

Students'
Physics
2.4 Engines produce polluting exhaust gases
2.8 Pollution and environmental problems of different energy sources
2.10 Air pollution in the home
3.18 Noise
4.1 Electrostatic attraction, used to remove ash from chimney gases
4.22 Thermal power stations produce waste gases
4.23 Acid rain; pollution and environmental problems of power generation
5.2 Dangers of nuclear radiation
5.3 Dangers from radioactive waste
Chemistry
4.3 Biodegradable plastics
6.4 Combustion reactions
13.4 Gases in the air
11.1 Oxides of nitrogen as pollutants
11.2 Ammonia and nitric acid in industry
11.3 Fertilizers
11.4 Fertilizer factory
11.5 Incomplete combustion of fuels produces carbon monoxide
11.7 Sulphur dioxide as a pollutant; acid rain
12.4 Pollution from burning fossil fuels

Teacher's Guide
Biology
A37 Pollution
A38 More pollution
Chemistry
A36 Ozone – a saver of life

9.9   More about acid rain

Topics covered: sources of air pollution; types of pollutant; how pollution spreads; dry and wet deposition; acid rain and its prevention.

Most air pollution is caused by the burning of fossil fuels. It is spread globally by convection currents (winds) in the atmosphere.

Polluting gases include oxides of sulphur and nitrogen. These dissolve in rain and turn it acid. Acid rain damages plants, buildings, soil and water life. Ways of tackling the problem include reducing emission levels from power stations and other sources, and reducing the demand for fuel-derived energy.

9.10   The greenhouse effect

Topics covered: the greenhouse effect; causes and effects of global warming; greenhouse gases; stopping the greenhouse effect.

The natural processes of photosynthesis and respiration keep the amounts of carbon dioxide and oxygen in the atmosphere in balance. However, the excessive burning of fossil fuels is upsetting the balance. Carbon dioxide levels are rising and the atmosphere is warming as a result. Ways of tackling the problem include reducing emission levels from power stations and other sources, and reducing the demand for fuel-derived energy.

9.11   Controlling pollution

Topics covered: energy conservation; alternative sources of energy.

All devices for converting one form of energy to another are inefficient. Internal combustion engines waste a great deal of energy as heat.

The burning of fossil fuels in engines and in power stations produces polluting gases. Catalytic converters can control the exhaust emissions from car engines. Some alternative methods of generating power do not produce gaseous pollutants.

9.12 Don't throw it away!

Topics covered: recycling waste materials; why recycling is important; renewable sources of energy.

Excessive use of packaging is wasteful and causes pollution. With the Earth's energy and other resources being depleted, it makes sense to recycle materials wherever possible. Metals are already extensively recycled because of their high cost.

The search is on for viable, renewable sources of energy. Waste materials can themselves be a source of energy.

9.13 Clean water and sewage disposal

Topics covered: pure water supplies; sewage disposal.

Reservoirs, created for water supply purposes, can also be used as a source of hydroelectric power.

The sewage treatment process relies partly on filtration and partly on bacterial action. The sludge from a sewage works can be used to make methane gas and fertilizers.

9.14 Soil

Topics covered: soil formed by weathering; separating out soil; fertile soil; fertilizers; life in the soil.

Soil starts to form when rocks are weathered. Weathering agents include frost, expansion and contraction, and the effects of acid rain. Soil is a mixture of many materials. Some of these can be separated by allowing a soil/water mixture to settle.

Acids or alkalis are added to soils to neutralize them. Fertilizers are added to increase crop yields.

9.15 Conservation and ecology
9.16 More about ecology

Topics covered: wildlife habitats; the destruction of natural habitats; conserving wildlife; ecology; communities and populations; adaptations; the environment.

The commercial and industrial needs of society are often in sharp conflict with the need to preserve natural habitats.

On both a local and a global scale, human activity can upset the cycles within the Earth's finely balanced ecosystems.

9.17 How to study wildlife I
9.18 How to study wildlife II

Topics covered: collecting specimens; line and belt transects; sampling plant and animal life.

Many physical and chemical factors may influence the distribution of plants and animals in a habitat. For example, some plants thrive in acidic conditions whilst others prefer alkaline soils.

Pollution will also affect the range of organisms to be found in a particular habitat.

9.19   Managed ecosystems

Topics covered: farms as managed ecosystems; use of insecticides and herbicides; protecting crops from the weather; improving crops and livestock; factory farms.

The production of food has been increased by the use of modern farming methods.

Chemical and biological methods are used to control pests. Soils are improved by the use of organic and inorganic fertilizers: pH levels are controlled by the addition of lime.

Greenhouses provide physical protection from the elements. Artificial lighting and automated irrigation and feeding systems allow fruits and vegetables to be grown 'out of season'.

The production of food in factory farms under intensive farming methods raises moral issues.

10 Health and hygiene

10.1 Germs, disease and infection
10.2 Defences against disease
10.3 Sexually transmitted diseases

Topics covered: viruses, bacteria and fungi; the spread and prevention of infection; natural immunity; antibodies; artificial immunity and vaccination; sexually transmitted diseases; syphilis; gonorrhoea; AIDS; transmission and symptoms.

Bacteria are single cells. They are about ten times larger than viruses. Viruses cause disease. So do some bacteria. Either may be spread by diffusion through the air.

Food poisoning is caused by bacteria. Bacteria in food can be killed by boiling, by using preservatives such as sulphur dioxide, or by irradiation. The reproduction rate of bacteria can be reduced or halted by chilling or freezing food.

Not all bacteria are harmful. Bacteria have an important part to play in supplying plants with nitrogen and returning nitrogen to the atmosphere. Bacteria are also used to treat raw sewage.

Students'
Physics
1.1   Relative size; size of cells
2.11 Diffusion
2.12 Temperature
5.3   Irradiation of food; nuclear radiation kills bacteria
Chemistry
1.1   Diffusion
1.4   Diffusion in gases
2.9   Radioactivity – killing microbes
11.7 Sulphur dioxide kills bacteria and moulds
13.7 Using bacteria – sewage treatment
13.6 Bacteria use dissolved oxygen in water
13.4 Nitrogen-fixing bacteria in root nodules; denitrifying bacteria in soil

Teacher's Guide
Biology
A44 Microbiology: safety precautions and techniques
A45 Growing and studying bacteria
A46 The effects of soap and pH on bacteria
A48 The effects of temperature on bacteria
I11   How safe to refrigerate?

10.4 Cancer

Topics covered: cancer; main causes of cancer.

Many cancers have been directly linked with controllable environmental factors. Some pollutants are known to be cancer-inducing, the most notable being cigarette smoke and radioactive contaminants. The Sun's ultraviolet radiation is known to cause skin cancer. There has been speculation about possible harmful effects from the alternating electromagnetic fields around power lines.

X-ray and gamma rays are harmful to living cells and have been successfully used to destroy malignant tumours. Increasingly, lasers are being used in certain types of cancer surgery.

Students'
Physics
2.10 Pollutants in a room
3.1   Lasers
3.10 Ultraviolet, X-rays and gamma rays
4.22 Overhead power cables
5.2   Dangers of nuclear radiation
5.3   Radioactive pollutants
5.4   Radiation used for cancer treatments
5.5   Hazardous radioactive decay products
Chemistry
7.6   Damaging the ozone layer
13.3 The ozone layer protects
13.4 Air pollution – damaging the ozone layer
14.2 Granite – some gives off radioactive gas

Teacher's Guide
Chemistry **A36** Ozone – a saver of life

10.5 Smoking and ill-health

Topics covered: poisons in tobacco smoke; diseases caused by smoking; passive smoking.

Cigarette smoke endangers other people's health as well as the smoker's. It quickly spreads by diffusion to fill any closed space. It contains poisonous carbon monoxide and ammonia, as well as 17 known carcinogens.

Students'
Physics **2.10** The need for air changes in a room **2.11** Diffusion **Chemistry** **1.1** Diffusion **1.4** Diffusion in gases **11.2** Ammonia **12.4** Effect of carbon monoxide on the blood **13.4** Air pollution

Teacher's Guide
Biology **A26** Lung volume **I6** Fitness and lung volume **T6** Smoking

10.6 Health warning

Topics covered: drugs and their effects; solvent abuse; drug abuse.

Solvent abuse is a growing problem. Solvents can cause sickness, depression and damage to the heart. In a sealed room, people can become unknowing 'solvent abusers' when they inhale the fumes from furniture, carpets and foam-filled walls.

Excessive intake of alcoholic drinks is the most widespread form of drug abuse.

Students'
Physics **2.10** Fumes from chipboard; the need for air changes in a room **Chemistry** **1.5** Solvents **12.8** Alcohol in drinks

Teacher's Guide
Biology **I10** Investigating fermentation

Teacher's Notes

This section contains 40 photocopiable 'activity sheets'. Most of these are designed to guide practical activities. Others are pen and paper activities which may be used in the class or as homework tasks.

Where practical tasks offer the opportunity to assess certain experimental skills, these are listed on the activity sheet. However, 'check boxes' are also provided so that the teacher can agree with the students which of the listed skills are to be assessed on each occasion. These may, of course, be the same for all students or may be negotiated with individuals.

Some of the activity sheets ask students to design investigations into scientific problems. They may find this easier if they have already been given the notes on investigations given in PART C of this guide.

Science 1: Each of the Activities detailed here provides ample opportunity for the practice and assessment of Sc1. Each Activity has been coded P, O, A, or E to assist teachers in identifying appropriate Sc1 opportunities To help further with this task each of the check boxes contains an open section which the teacher can use to define precisely the specific aspect of Sc1 he or she wishes a student to focus on. (Note: most *planning* (P) opportunities occur in the Investigations section in PART C of this guide.)

P – planning experimental procedures

O – obtaining evidence

A – analysing evidence and drawing conclusions

E – evaluating evidence

Activity	HAZARD WARNING	P	O	A	E
A1			✔		
A2	✔		✔	✔	
A3	✔		✔	✔	
A4	✔		✔		
A5			✔	✔	✔
A6			✔	✔	✔
A7	✔			✔	
A8			✔		
A9	✔		✔		
A10	✔		✔	✔	✔
A11	✔		✔	✔	
A12	✔		✔	✔	
A13			✔	✔	✔
A14			✔	✔	
A15			✔	✔	✔
A16	✔		✔	✔	✔
A17			✔	✔	
A18	✔		✔	✔	
A19	✔		✔	✔	
A20	✔	✔	✔	✔	✔
A21			✔		
A22	✔		✔	✔	✔
A23	✔		✔	✔	
A24	✔		✔	✔	

Activity	HAZARD WARNING	P	O	A	E
A25	✔		✔	✔	
A26		(✔)		✔	
A27			✔	✔	
A28	✔		✔	✔	✔
A29	✔		✔	✔	✔
A30	✔		✔	✔	✔
A31	✔		✔	✔	
A32			✔	✔	
A33			✔	✔	
A34	✔		✔		
A35			✔		
A36			✔	✔	
A37			✔	✔	✔
A38			✔	✔	
A39	✔		✔	✔	
A40	✔		✔	✔	
A41	✔		✔	✔	
A42			✔	✔	
A43	✔		✔	✔	✔
A44		–	–	–	–
A45	✔		✔	✔	
A46	✔		✔	✔	
A47	✔		✔		
A48	✔		✔	✔	

Students' Notes: Getting started ...

The aim of this book is to help you learn the techniques and skills which biologists use to study the living world. When you are doing experiments remember:

Be safe

Some important do's . . .

Do keep bags and coats well away from the work area, where they won't cause accidents.

Do put thermometers where they can't roll off the bench.

Do take care of expensive apparatus.

Do leave bunsens on a yellow flame when they aren't being used, so that the flame can be seen.

Do leave hot tripods and beakers to cool down before moving them.

Be skilful

Learn and practise the basic laboratory techniques including:

- how to use a microscope
- how to use dissecting instruments
- how to perform chemical tests on foods and other materials
- how to investigate living things in their natural environments

Be accurate

When using a ruler: be sure that the scale is right alongside the points you are trying to measure.

When measuring the temperature of a liquid: keep the liquid well stirred, give the thermometer time to reach the temperature, keep the bulb of the thermometer in the liquid while you take your reading.

When measuring a liquid level on a scale: Use the level of the flat surface, not the edge of the meniscus.

When taking readings: Take plenty! For a graph, you should take at least five sets of readings. For a single measurement, repeat at least three times and find an average value.

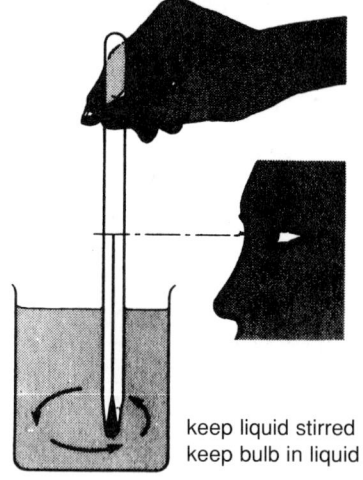

keep liquid stirred
keep bulb in liquid

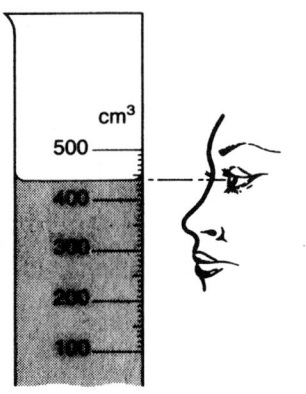

measure to flat
surface of liquid

Record your readings

Draw a results table for your readings *before* you start your experiment. Remember to write down exactly what you are measuring and the units you are using. Put headings at the top of any columns you use.

When you take readings, put them directly in your table, *not* on a piece of scrap paper.

Write a report

Keep it brief. Just write down:

1 the purpose of the experiment
2 what you did, and the order in which you did it
3 what measurements you had to take
4 any special precautions you took – for example, covering the soil in a pot with a plastic bag so that water could not evaporate from it
5 any calculations you did

Draw a graph

If you have several sets of readings, you can draw a graph.

Decide which readings are going to go along the bottom axis: Usually it is the readings you had control over. For example, if you decided to read the temperature every minute, then time (in minutes) goes along the bottom axis.

Choose the largest scales you can for your axes: For example, 2 squares for every minute is better than 1 square for every minute – provided that all the readings will fit on the graph paper. Check your highest readings before you decide on a scale.

Label your axes: Each should show what is being measured and the unit used.

Plot your points: Most people use a small cross to mark each point.

Draw the best line you can through the points: Experimental readings aren't exact, so the points will probably zig-zag a little. Don't join up the points. Decide whether the line should go through the origin. Then draw the straight line or smooth curve which goes through most of the points. If you've got it right, there should be roughly as many points on one side of your line as the other.

Write down your conclusions

You've done the experiment. You've written a report and plotted a graph. Finish off by writing down your conclusions – what you've found from the experiment.

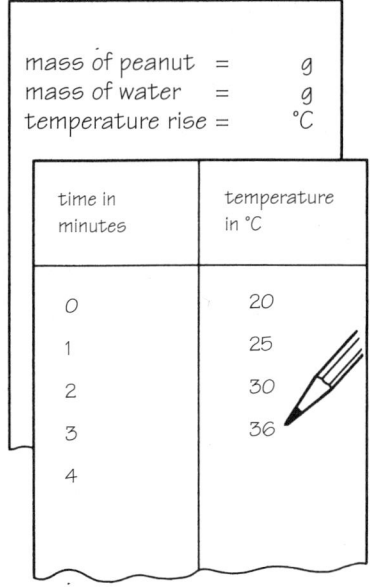

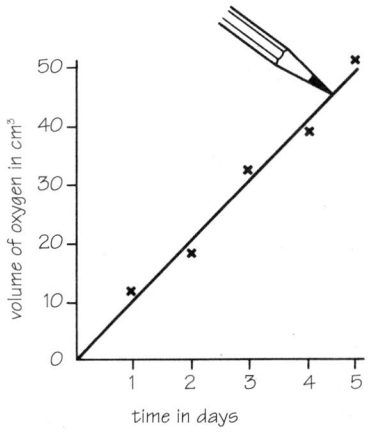

Microscopes and how to use them	**BIOLOGY** _ACTIVITY_	**A1**

You need:

a microscope
prepared slides (e.g. insect head, wings, legs, etc.)

The parts of a microscope

Coarse focus knob This is used to get the specimen roughly in focus.

Fine focus knob This is used to get the specimen in sharp focus.

Iris adjuster This controls the amount of light reaching the specimen.

Light switch For the bulb used to shine light through the specimen.

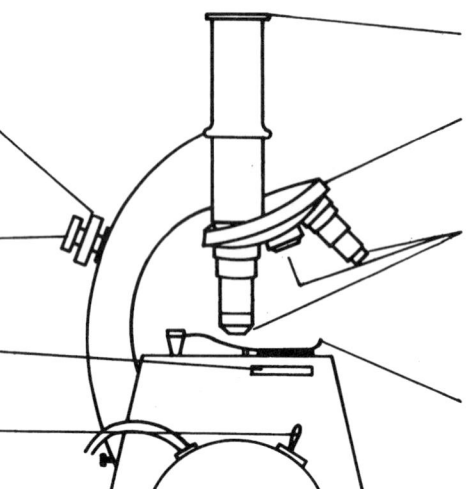

Eyepiece This is the lens you look through.

Turret This turns round so you can use different objective lenses.

Objective lenses These have different powers of magnification. The longer the lens the greater its power to magnify specimens.

Stage The platform that you put the specimen slides on. Slides are held down by a spring clip.

Using a microscope

1 Turn the turret until you have the low power objective lens (the short lens) in line with the eyepiece.

2 Clip a slide on the stage so that it is in the centre under the objective lens and look through the eyepiece.

3 Adjust the coarse focus until the specimen becomes clear. if necessary adjust the fine focus until the specimen is in sharp focus.

4 Move the iris adjuster until the specimen is clearly lit.

5 Calculate the magnification by multiplying the power of the eyepiece by the power of the objective lens (e.g. a × 5 eyepiece used with a × 15 objective magnifies 75 times).

6 Notice how, when you move the slide, the specimen seems to move in the opposite direction.

7 Change to the medium power objective. **Do not use the higher power objective yet.** Focus the microscope and notice that you now see much less of the specimen but at a higher magnification.

8 **How to use high power objectives.** If this is done carelessly the lens and a slide can be damaged.

   a) With your eyes level with the stage, slowly lower the high power objective until it *almost* touches the slide.

   b) Look through the eyepiece and focus by moving the lens *away from the slide* (i.e. always focus upwards). This avoids smashing the lens through a slide.

Your teacher will be looking for:	**and especially for:**
• *careful* use of the apparatus given	☐
• good observation of the point where the image is in sharp focus	☐
•	☐

Making microscope slides

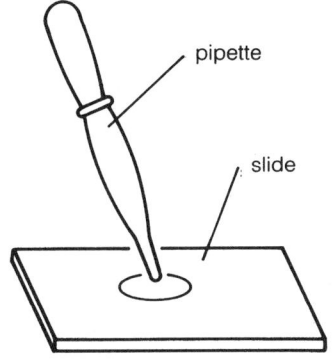

You need:

microscopes

mounted needles

slides and coverslips

scissors

pond or aquarium water

newspaper with words and pictures

crystals (salt, sugar, potassium manganate(VII), copper sulphate)

dropper pipettes

Pond and aquarium water

Water from the bottom of a pond or aquarium, especially if it contains rotting vegetation, can contain many different protozoa and other microscopic creatures.

1 Use a bulb pipette to place one drop of pond or aquarium water onto the centre of a glass slide.

2 Place a coverslip with one edge resting on the slide near the drop of water. Use a mounted needle to lower it slowly onto the water. If you do this quickly you will trap air bubbles. Use just enough water to spread to the edges of the coverslip and no further. Place the slide on the microscope stage.

3 Start with low power magnification and search the slide for interesting objects, then change to medium or high power magnification.

4 Make notes and drawings of what you find.

More things to do

5 Put a drop of tap water onto a slide. Remove a hair from your head, place it across the water and lower a coverslip over it. Study it under medium and high power magnification and make notes and drawings of the root end, the middle and the upper end of the hair.

6 Cut out pieces of newspaper small enough to fit under a coverslip. Mount them in water on slides. What is a newspaper photograph made up of?

7 Sprinkle some crystals on a *dry* slide. Study them without a coverslip. Prepare a table and use drawings and words to compare the shape and colour of four different types of crystal.

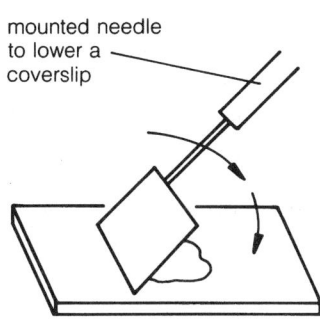

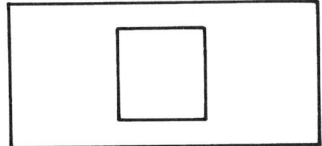

Correctly prepared slide

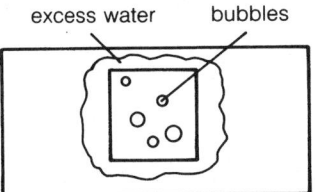

Badly prepared slide

HAZARD WARNING

Copper sulphate is harmful when swallowed. It may also be irritating to eyes and skin. Potassium manganate VII is harmful if swallowed. AVOID SKIN CONTACT. WEAR EYE PROTECTION.

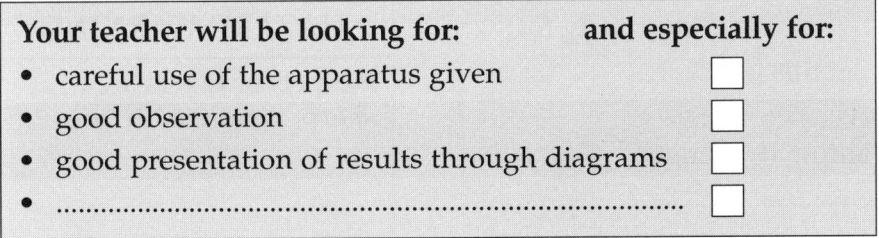

Your teacher will be looking for:	and especially for:
• careful use of the apparatus given	☐
• good observation	☐
• good presentation of results through diagrams	☐
•	☐

Looking at cells

You need:

microscopes	forceps
razors or scalpels	onions
slides and coverslips	kidneys
Petri dishes	moss plants

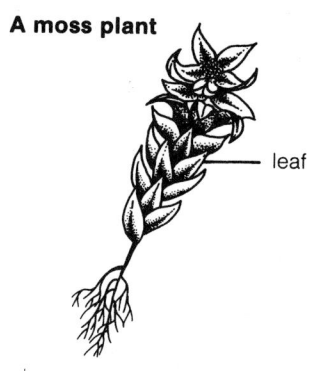

A moss plant

leaf

Moss leaf cells

1 Use forceps to take one leaf off a moss plant. Put the leaf on a slide, add a drop of water and lower a coverslip onto it.

2 Observe it under low, medium and high power. Identify as many parts as you can.

Onion cells

3 If you look at half an onion, you will see that it is made of fleshy leaves. Use a razor to cut a small piece out of one of the leaves. Use forceps to peel skin off the *inner* surface of the leaf. This skin is a thin layer of living cells. Put the skin into a Petri dish of water.

4 Put a drop of iodine stain onto a slide. Put a piece of onion skin into the stain and smooth it out so there are no folds. Lower a coverslip over it, taking care not to trap any bubbles. Prepare another slide in the same way but using water instead of iodine stain.

5 Study the stained onion cells under different magnifications, then look at unstained cells. What parts of the cells have become stained? How are onion cells *different from,* and *similar to,* moss leaf cells?

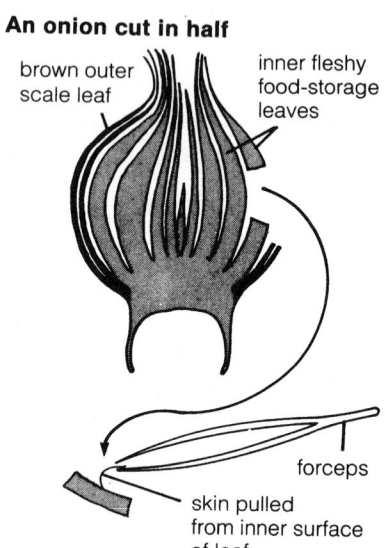

An onion cut in half

brown outer scale leaf

inner fleshy food-storage leaves

forceps

skin pulled from inner surface of leaf

Animal cells

6 Use a razor and forceps to peel small pieces of transparent skin off the outside of a kidney. Make a slide of the skin in water, and another in iodine.

7 Study stained and unstained cells. How are they different? Draw moss leaf, onion and animal cells and list their *similarities* and *differences*.

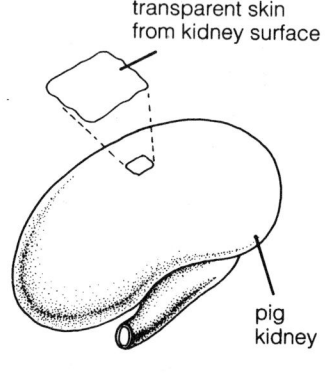

transparent skin from kidney surface

pig kidney

Your teacher will be looking for:	and especially for:
• careful and skilful use of the apparatus given	☐
• good observation of cell structure	☐
• good presentation of results including drawings	☐
•	☐

HAZARD WARNING

Scalpels or razors are sharp, handle with care.

Measuring cells	BIOLOGY ACTIVITY	A4

You need:

microscopes

slides and coverslips

razor blades

Petri dishes

onions and moss plants

clear plastic rulers

salt and other crystals

insect slides (permanent)

Measuring a field of view

1 Place a clear plastic ruler under a microscope and focus on it with low power magnification. How many millimetres wide is the field of view?

2 **Problem:** Microscopic objects are measured in **micrometres** (one micrometre is written 1 μm). 1 mm = 1000 μm. Convert your field of view to micrometres.

Measuring onion cells

3 Prepare a slide of onion cells. Look at the slide under low power magnification. How many cells fit across the field of vision? In the drawing opposite, four and a half cells fill a field of view 2200 μm wide. What is the average length of each cell?

4 What is the average length, in micrometres, of onion cells in your slide? Turn the slide around and calculate the average width of the cells.

5 You now know the length in micrometres of one onion cell. Use this information, and your onion slide, to calculate the field of vision in micrometres under medium and high power magnification.

More things to do

6 Using the technique you have learned, measure:

a) the length and width of moss leaf cells
b) the width of a human hair
c) the average size of sugar, salt and other crystals

7 Look at permanent slides of insects and measure various parts, such as the width of scales on a butterfly's wing, the width of lenses in an insect's compound eye, the size of a fly's foot, etc.

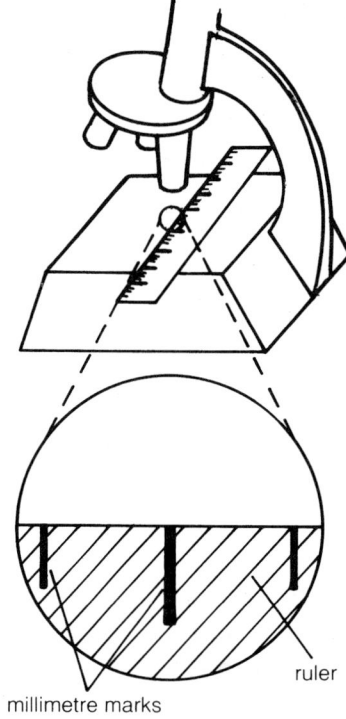

millimetre marks · ruler

The field of view is 2.2 mm.
What is this in micrometres?

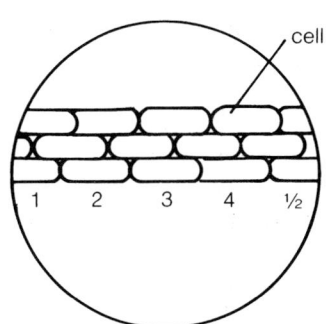

The field of view is 2.2 mm.
What is the length of one cell, in micrometres?

Your teacher will be looking for:	and especially for:
• careful use of the apparatus given	☐
• accurate measurements and calculations	☐
• good presentation of results	☐
•	☐

HAZARD WARNING

Razor blades are sharp, handle with care.

Looking at variation	BIOLOGY ACTIVITY	A5

You need:

tape measure

scales (in kilograms)

ruler with millimetre scale

graph paper

1 Working in pairs, record the following information for both partners (refer to drawings below if necessary).

   a) Are your ears lobed or unlobed (drawing A)?
   b) Can you roll your tongue (drawing B)?
   c) What is the length of your little fingers, to the nearest millimetre (drawing C)?
   d) What is your mass, in kilograms?

name	ear lobes yes/no	tongue rolling yes/no	finger length (mm)	mass (kg)

2 Record your results

   a) Data for the whole class should be recorded in the form of a table on the chalkboard (see opposite).
   b) Data should be used to draw histograms. A blank histogram for finger length is shown opposite. Draw others for the remaining data.

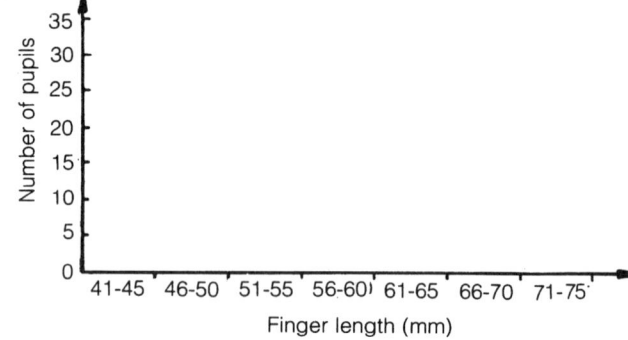

3 Use your results to decide which of the characteristics shows **continuous variation** and which shows **discontinuous variation**.

A Ears

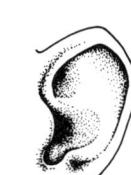

ear lobe present   ear lobe absent

B Tongue

rollers or non-rollers

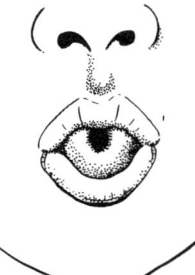
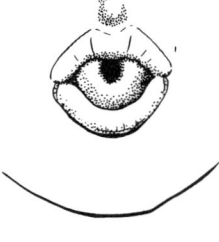

C Finger length (mm)

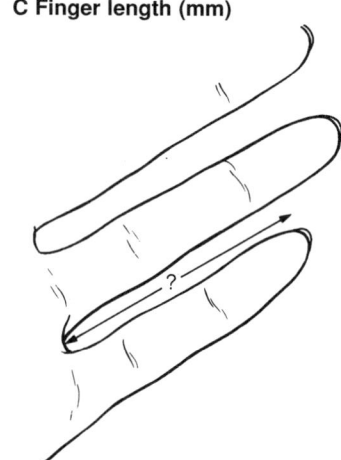

Your teacher will be looking for: **and especially for:**

• careful measurement and recording of results ☐

• good presentation of results in tables and histograms ☐

• .... ☐

Further work

   a) Think of other variations you could investigate for humans. For each one, predict whether it is continuous or discontinuous.
   b) Think of variations you could study in plants. Investigate one of these by collecting data and presenting your results as a histogram.

A model for genetics

You need:

two-pence coins

circular sticky labels which fit a coin

H and **h** are alleles of the gene for hair colour.

H is dominant and if it is present the hair is dark. If both alleles are **h** the hair is fair.

ova

		H	h
sperms	H	HH	Hh
	h	hH	hh

Look at the diagram above showing the zygotes which could be produced by a father with the genotype **Hh** and a mother with the genotype **Hh**.

The diagram shows that zygotes with a **dominant allele (HH, Hh or hH)** are three times more likely to be produced than zygotes with two **recessive alleles (hh)**. In other words, dominant and recessive phenotypes occur in the ratio of 3:1.

This happens because:

- half the sperms carry the **H** allele and half carry the **h** allele
- half the ova carry the **H** allele and half carry the **h** allele
- there is an equal chance that, during fertilization, any sperm can fertilize any ovum (i.e. fertilization is a **random process**)

This can be checked experimentally using coins to represent sperms and ova, and coin tossing to represent the random process of fertilization.

1 Work in groups of two. Each group should obtain two 2p coins. Use circular sticky labels to mark one side of one coin **sperm A** and the opposite side **sperm a**, then label one side of the other coin **ovum A** and the opposite side **ovum a**. Copy the chart below.

2 Working in pairs, spin the 'sperm coin' and the 'ovum coin' at the same time. Look at how they fall and enter the result in the appropriate part of the tally column. Repeat at least 50 times, then work out the totals.

3 Does the ratio of dominant to recessive phenotypes come to about 3:1?
Why must the coins be tossed at least 50 times to get reliable results?

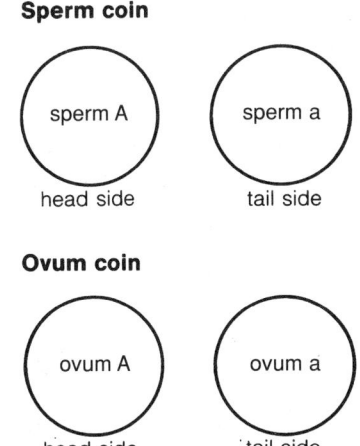

Sperm coin

sperm A — head side   sperm a — tail side

Ovum coin

ovum A — head side   ovum a — tail side

sperm	ovum	tally
A	A	
a	A	
A	a	
a	a	

} total =

} total =

Further work

B and **b** are the alleles for eye colour. **B** is dominant and if present the eyes are brown. If both alleles are **b** the eyes are blue. A blue eyed man has the genotype **bb**. He is married to a brown eyed woman with the genotype **Bb**.

a) What are the chances that their first child will have blue eyes?

b) What are the chances that their second child will have blue eyes?

c) If they have four children, how many will have brown eyes? (Be careful – can you be **sure**?)

An alcohol problem

Alcohol made from plants is a very useful fuel. In some countries, including Brazil, it offers a good alternative to petrol made from oil. The diagram shows the main stages in making alcohol for fuel.

HAZARD WARNING

Ethanol is highly flammable. KEEP AWAY from naked flame. WEAR EYE PROTECTION.

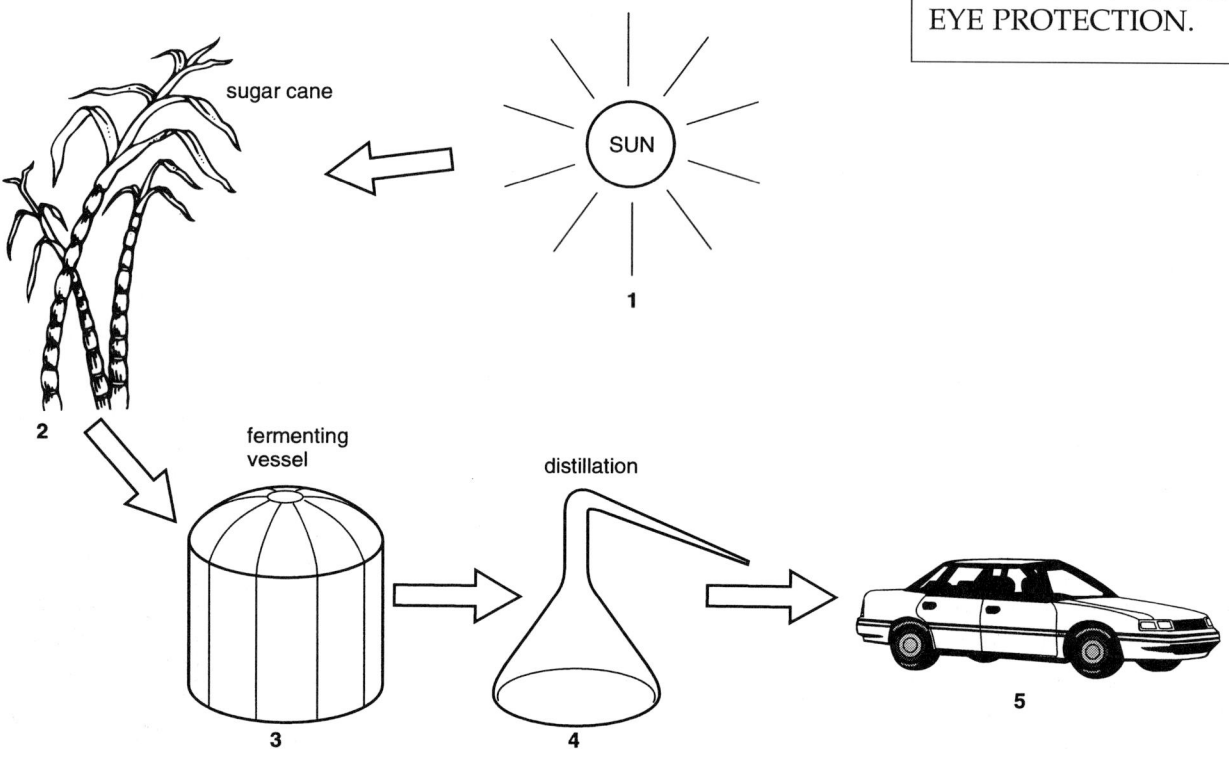

1 What form of energy given off by the sun is used by plants?

2 What process is going on in the sugar cane plant to make sugar?

3a The sugary sap can be fermented. What will happen to the sugar?

3b What gas is given off during fermentation? Give one use for this gas.

4 The end result of fermentation is a mixture of alcohol and water. How will distillation be used to separate the two liquids?

Further work

a) Brazil would like to replace all its petrol with alcohol fuel. Suggest some reasons why.

b) About 50% of Brazil's agricultural land would have to be planted with sugar cane to produce enough alcohol. What problems might this cause?

c) What could Brazil do to maintain food stocks for its people?

d) Brazil is covered in part by the Amazon Jungle, a tropical rain forest. The Brazilians may be tempted to cut down large parts of the forest for use as agricultural land. What effects might this have on the balance of gases in our atmosphere? Explain your answer.

Photosynthesis and oxygen

You need:

a litre beaker	Bunsen burner
glass funnel	bench lamp
test tube	sodium hydrogen carbonate
Plasticine	spatula
wood splint	pond weed *(Elodea)*

1 Three-quarters fill a 1 litre beaker with water in which a small amount of sodium hydrogencarbonate has been dissolved. This will supply the plants with carbon dioxide.

2 Put a few springs of healthy pond weed such as *Elodea* in the bottom of the beaker. Place a glass funnel over the pond weed. Use one or more lumps of Plasticine to raise the rim of the funnel off the base of the beaker, so the liquid can circulate freely. Make sure the liquid level is well above the end of the funnel.

3 Fill a test tube with weak sodium hydrogen carbonate solution. Put your thumb over the end, turn the tube upside down and lower it into the beaker without letting in any air. When the end of the test tube is under the liquid remove your thumb and lower the tube onto the funnel, as shown in the diagram.

4 Either put the apparatus on a well-lit window ledge or place it near a bench lamp.

5 After about a week sufficient gas should have collected in the tube to test.

Lift the test tube off the funnel but do not let in any air. Put your thumb over the end of the tube, lift it out of the beaker and turn it right way up. Do not remove your thumb yet.

Test the gas for the presence of oxygen: light a wood splint and when it is burning brightly blow it out so the end is glowing red hot. Lift your thumb off the test tube and *very quickly* lower the glowing wood splint into it. Observe closely what happens.

6 **Questions:**
Does the tube contain pure oxygen?
Can you devise a control for this experiment?

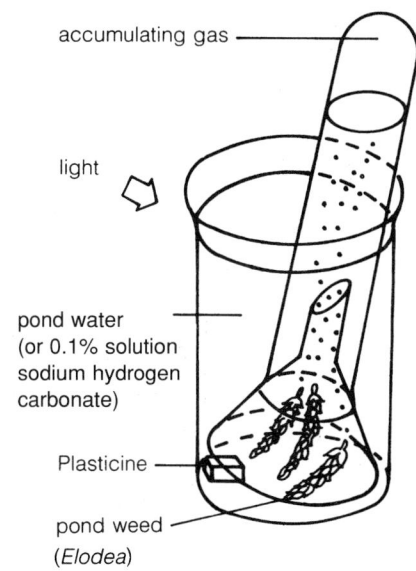

accumulating gas

light

pond water (or 0.1% solution sodium hydrogen carbonate)

Plasticine

pond weed *(Elodea)*

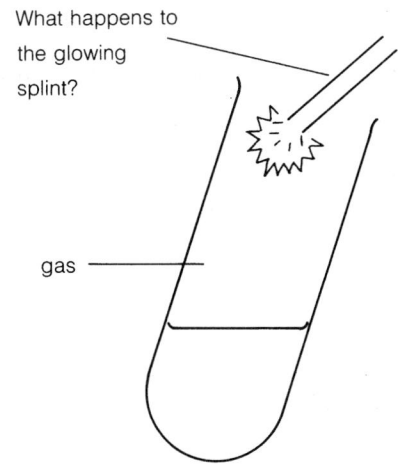

What happens to the glowing splint?

gas

Your teacher will be looking for:	and especially for:	
• careful use of the apparatus given		☐
• accurate observations		☐
• good presentation of results		☐
• sensible conclusions which fit your results		☐
•		☐

Chlorophyll and photosynthesis

You need:

plants with normal and variegated leaves (pelargonium, coleus or geranium)

iodine solution

boiling tubes

tripods and gauzes

forceps

500 cm³ beakers

white tiles

Bunsen burners

ethanol (alcohol)

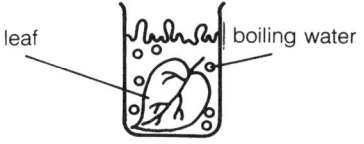
leaf — boiling water

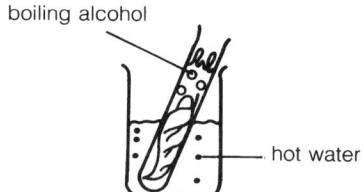
boiling alcohol — hot water

How can you show that plants need chlorophyll for photosynthesis?

You need a way of showing that photosynthesis has taken place. Plants change sugar produced by photosynthesis into starch and store it in their leaves. So a leaf with starch has been carrying out photosynthesis.

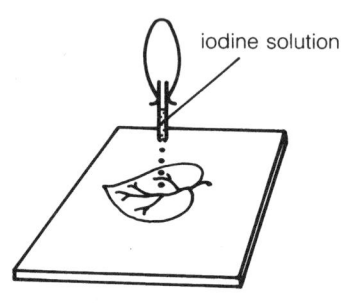
iodine solution

Testing a leaf for starch

1 Take a leaf from a non-variegated plant which has been in the light for a few hours. *SAFETY: put goggles on.* Half fill a 500 cm³ beaker with water and bring it to the boil. Put the leaf in the water for about 1 minute, then turn the Bunsen off.

2 Half fill a boiling tube with ethanol. *SAFETY: this is highly inflammable so do not put it near a naked flame.* Use forceps to take the boiled leaf out of the water and transfer it to the ethanol. Put the tube of ethanol into the beaker of very hot water. The ethanol will boil and remove chlorophyll from the leaf, making test results easier to see.

3 Lift the leaf out of the ethanol, dip it into the hot water to soften it, spread it out on a white tile and cover it with iodine solution. A blue-black colour indicates the presence of starch in the leaf.

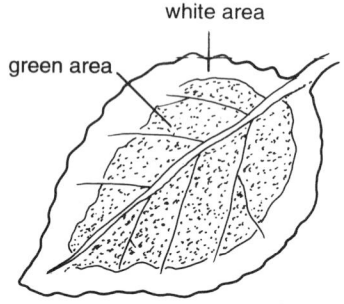
A variegated pelargonium leaf
white area
green area

Your teacher will be looking for:	and especially for:	
• careful and safe use of the apparatus given		☐
• accurate observations		☐
• good presentation of results		☐
• sensible conclusions which fit your results		☐
•		☐

? You are provided with variegated leaves – leaves which have areas with and without chlorophyll. Design an experiment using these leaves to show that chlorophyll is necessary for photosynthesis.

<table>
<tr><td>

Light, carbon dioxide, and photosynthesis

</td><td>

BIOLOGY
ACTIVITY

</td><td>

A10

</td></tr>
</table>

You need:

potted plants	iodine solution
scissors	paper clips
ethanol	soda lime
white tiles	conical flasks
boiling tubes	clamps and stands
500 cm^3 beakers	cotton wool
Bunsens, tripods and gauzes	Vaseline
black paper or polythene	spatulas

Do plants need light for photosynthesis?

1 You could test two plants for starch – one which had been in the dark for 12 hours and one which had been in the light. But perhaps you can think of more interesting experiments using the materials provided?

2 How could you use strips of black paper or polythene, or even a black and white 35 mm photographic negative?

3 At which stage will you detach the experimental leaf from the plant?

4 Predict what a leaf will look like after the starch test. Make drawings of your results.

Find out: Do plants need carbon dioxide for photosynthesis?

You could use the apparatus in the drawing opposite.
Why must you start with a plant which has been in the dark for 12 hours?
Why must you test a leaf for starch *before* setting up the apparatus opposite?
Find out what the soda lime will do to air in the flask.
Why is the Vaseline necessary?
What control is needed?
Where will you put the plant and for how long?
What test is needed to obtain a result?
How will you present results and conclusions?

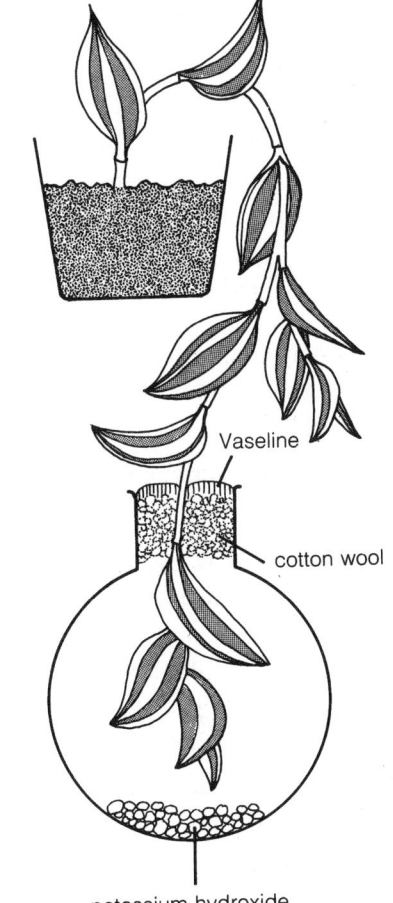

Vaseline

cotton wool

potassium hydroxide

HAZARD WARNING

Ethanol is highly flammable. KEEP AWAY from naked flame. Soda lime (sodium hydroxide and calcium hydroxide) is CORROSIVE and can cause severe burns; also dangerous to eyes and skin. AVOID SKIN CONTACT. AVOID CONTACT WITH WATER. WEAR EYE PROTECTION.

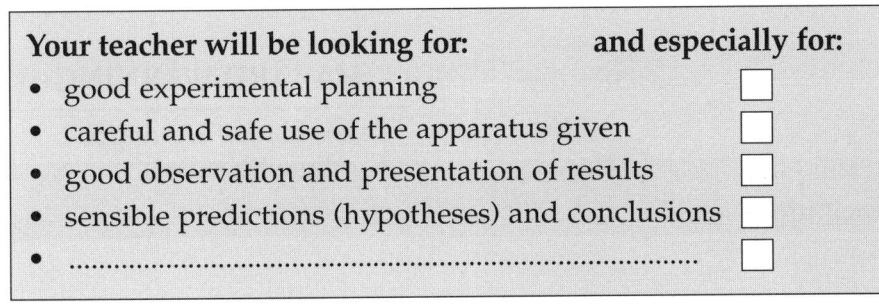

Your teacher will be looking for: **and especially for:**

- good experimental planning ☐
- careful and safe use of the apparatus given ☐
- good observation and presentation of results ☐
- sensible predictions (hypotheses) and conclusions ☐
- .... ☐

Transport tissue in plants

You need:

microscopes	white tiles
slides and coverslips	Petri dishes
razors	paint brushes
celery	eosin dye
germinated broad beans	

Water-conducting tissue of celery

1 Obtain a stick of celery, preferably with leaves still attached. Put it in a beaker half-filled with eosin dye and leave it for 24 hours.

2 Look carefully at the leaf veins. Observe and describe what has happened. Explain what has happened.

3 Lay the celery on a white tile and use a razor to cut thin slices off it. Continue until you have a slice so thin it is almost transparent.

4 Use a paint brush to transfer the slice to a microscope slide, add a drop of water and lower a coverslip over it.

5 Make a drawing of the slide showing which areas have turned red. What are these areas? Refer to page 66 of the text book.

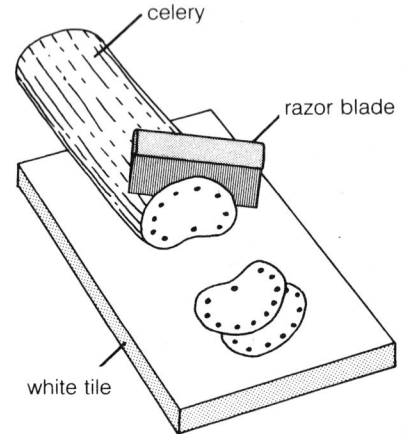

Compare root and stem of beans

6 Germinate a number of broad bean seeds by trapping them against the sides of a jam jar with a cylinder of blotting paper filled with damp sand or sawdust. Leave them until the root and stem have developed.

7 Clamp a bean over a beaker of eosin so that its root is immersed in the dye. Leave it until the dye becomes visible in the leaf veins.

8 Cut thin slices of the root and stem, and make drawings to show which areas have been stained red.
   What is the difference between the position of xylem in a bean stem and root?

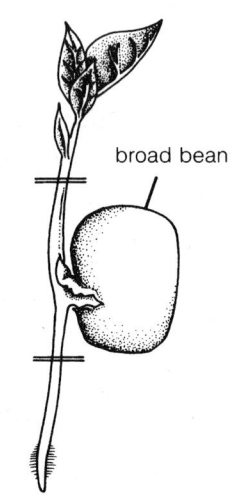

Your teacher will be looking for:	**and especially for:**
• careful use of the apparatus given	☐
• good observation	☐
• good presentation of results as diagrams	☐
•	☐

HAZARD WARNING

Razor blades are sharp, handle with care.

Osmosis

You need:

potatoes	scalpels
Petri dishes	sugar
dandelion stems	razor blades
test tubes and racks	

Osmosis in potato cells

1 Make three potato cups from raw potatoes cut in half: cut a depression in the top, peel skin off the sides, and give each a flat base (see diagram opposite).

2 Boil one cup. Place the cups in Petri dishes of water. Pour sugar into the boiled cup and into one of the raw potato cups. leave one cup empty.

3 Observe and describe what happens to the sugar, and what happens in the empty cup.
Explain what happens in the three cups.
Why was one cup left empty?
What conclusions can you draw about osmosis in living and dead cells?

Osmosis in dandelion stalks

4 Take two test tubes. Half-fill one with water and the other with strong sugar solution.

5 Obtain two dandelion stalks. Slit them upwards for about 2.5 cm, then make a second upward slit at right angles to the first to divide the stalk base into four strips (see diagram).

6 Put one stalk in water and the other in sugar solution and leave them for 10 minutes.

7 Observe and describe what happens.
What happens if you move the stalk in water to the sugar solution and vice versa?

A – Empty potato cup

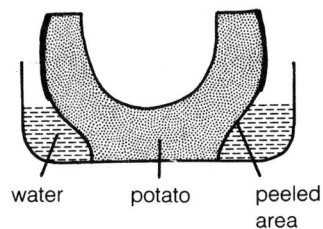

water   potato   peeled area

B – Raw potato cup with sugar

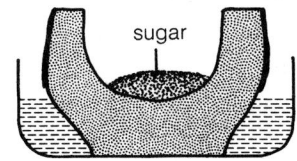

sugar

C – Boiled potato cup with sugar

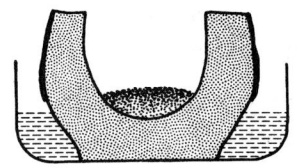

dandelion stalk

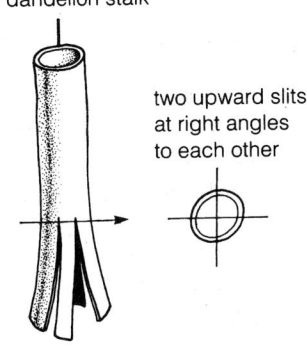

two upward slits at right angles to each other

Your teacher will be looking for: **and especially for:**

- skilful and safe use of the apparatus given ☐
- good observation of results ☐
- good presentation of results (including diagrams) ☐
- sensible conclusions ☐
- .... ☐

HAZARD WARNING

Razor blades are sharp, handle with care.

Further work

The diagram opposite shows a section through a dandelion stalk. Study it and use it to form a hypothesis to explain what you observed in your experiment.

Section through stalk

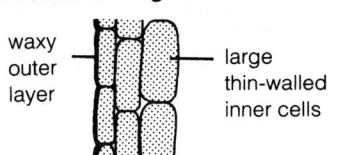

waxy outer layer   large thin-walled inner cells

Measuring transpiration

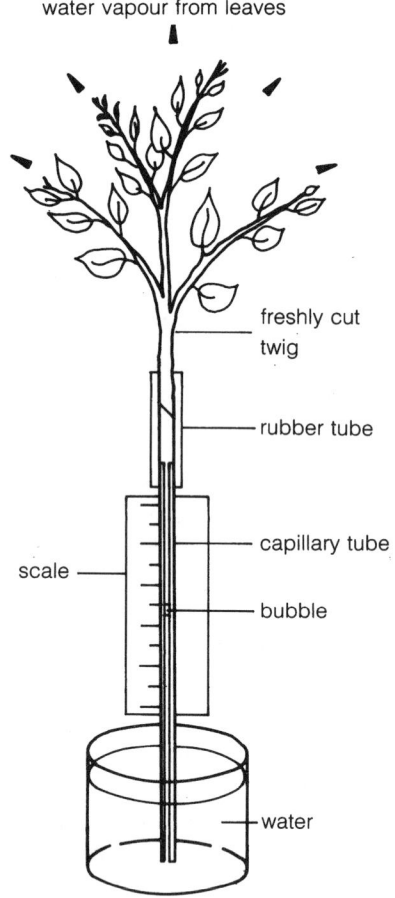

You need:

privet or other leafy twigs	cotton wool
capillary tubes	rubber tubing
Vaseline	labelled specimen tubes

Making a potometer

1 Set up the apparatus opposite in the following way. Push the piece of rubber tubing about 2 cm over the end of the capillary tube, submerge in water and squeeze the rubber tube until *both* tubes are full.
Push the end of the twig into the rubber tube (don't wet the leaves) and, without letting water escape, clamp the apparatus over a beaker of water. Seal all joints with Vaseline and fasten a card scale in position (see diagram).

2 After five minutes raise the apparatus so the capillary is out of the water. Air should start to move up the capillary. How far does it move in two minutes? Find the average time for three, two-minute runs.

3 What *exactly* have you measured?
Is what you measured the same as transpiration?
How could you measure the *volume* of water the twig takes up each minute?

Find out: The effects of climate on transpiration

Method one

Devise an experiment using the apparatus above to discover the effects on the twig of five *different* climatic conditions (listed in method two).

Method two

What *measurable* change will take place in the apparatus opposite with time? What will cause this change?

Design an experiment using this apparatus to discover the effects on transpiration of hot, cold, windy, humid and dark conditions.

Start by thinking about the following:
How you will create these conditions?
What will you measure, and when?
What controls are needed and how will you record your results?
Remember: one condition is changed; others remain the same.

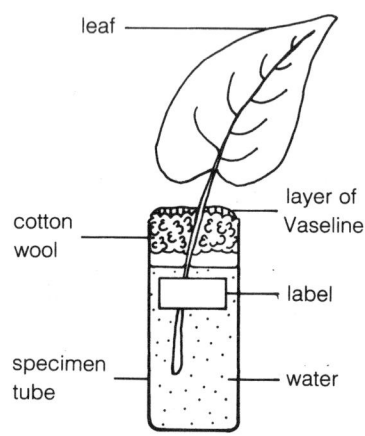

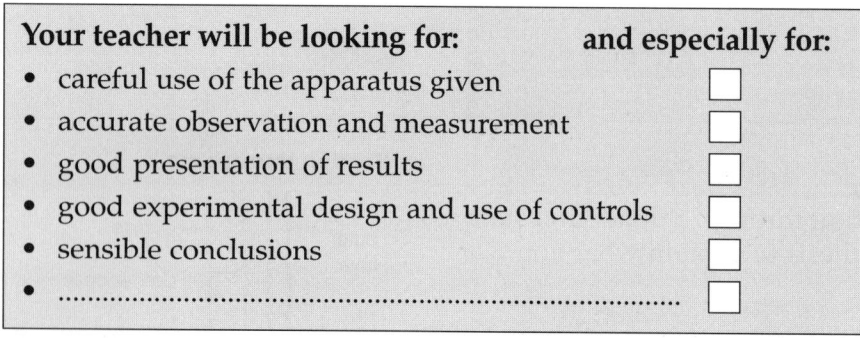

Your teacher will be looking for:	and especially for:
• careful use of the apparatus given	☐
• accurate observation and measurement	☐
• good presentation of results	☐
• good experimental design and use of controls	☐
• sensible conclusions	☐
•	☐

Stomata	BIOLOGY	A14
	ACTIVITY	

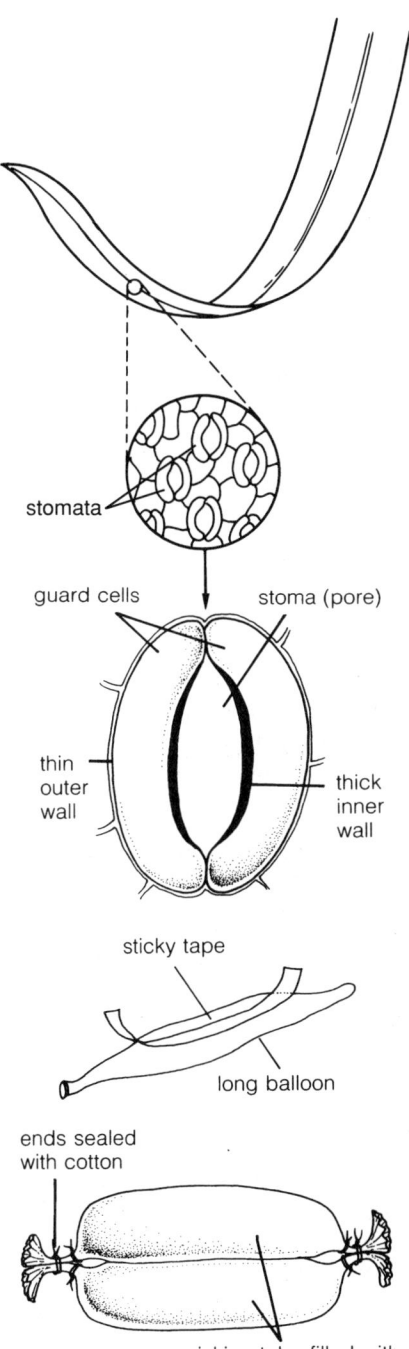

You need:

microscopes — long balloons

slides and coverslips — visking tubing

fine-point scissors — cotton thread

sugar — transparent sticky tape

iris leaves

Looking at stomata

1 It is possible to tear an iris leaf so that a small area of lower skin is exposed. Hold a leaf with its lower surface uppermost and tear it diagonally rather than straight across.

2 Make a microscope slide of lower skin and look at it under a magnification of at least × 600.

3 Estimate the size, in micrometres, of stomata.
Roughly how many stomata are there per square millimetre of leaf surface?

A balloon model of a guard cell

4 Fasten a length of sticky tape to one side of a long balloon. Inflate the balloon and compare its shape with another balloon without the sticky tape.

Problem: Guard cells have thicker walls next to the stoma than on their outer walls (diagram above). Think about the shape of the balloon with the sticky tape, and formulate an hypothesis to explain the function of a guard cell's thick inner wall. What effect could it have as guard cells inflate and deflate to open and close a stoma?

More things to do

Make a visking tube model of stomata.
Fill two 20 cm lengths of visking tube with strong sugar solution and seal both ends with cotton. Tie the two visking sausages firmly together, place them in water for 30 minutes and explain what happens.
What do the two visking sausages represent?
Use this result to formulate an hypothesis to explain how stomata open and close.

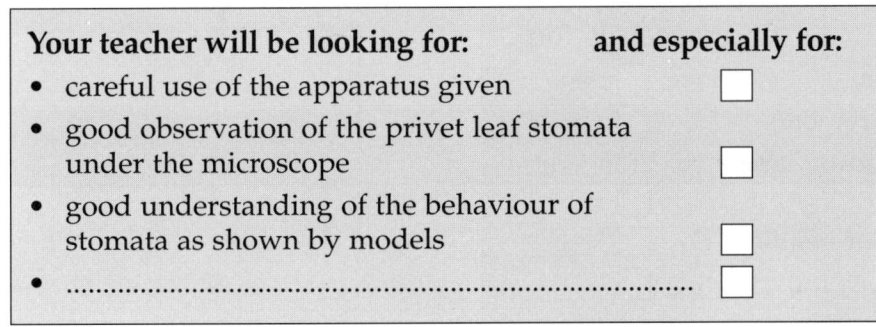

Your teacher will be looking for: **and especially for:**

- careful use of the apparatus given ☐
- good observation of the privet leaf stomata under the microscope ☐
- good understanding of the behaviour of stomata as shown by models ☐
- .... ☐

Germination and growth

You need:

cress and pea seeds	cotton wool
paper towels	test tubes and racks
500 cm³ beakers	paraffin oil
balances	refrigerator

1 warmth light air **2** warmth water air **3** cold water air **4** warmth water light air **5** warmth water light

Germination

1 Label five test tubes 1 to 5.

2 Put cress seeds in tube 1 and place it in a warm, well-lit place. These seeds have warmth, air and light but no water.

3 Put cress seeds on wet cotton wool in tube 2 and place it in a warm dark place. These seeds have warmth, air and water but no light.

4 Put cress seeds on wet cotton wool in tube 3 and place it in a refrigerator. These seeds have water and air but no warmth or light.

5 Put cress seeds on wet cotton wool in tube 4 and place it in a warm, well-lit place. These seeds have warmth, water, air and light.

6 Put cress seeds in tube 5 and cover them with boiled and cooled water (boiling drives oxygen out of the water), then pour a little paraffin oil onto the water to keep oxygen out. Place the tube in a warm, well-lit place. These seeds have warmth, water and light but no oxygen.

7 Examine the tubes after about three days. In which tubes have seeds germinated? What conditions are necessary for seeds to germinate?

Growth curves

If you weigh organisms as they grow and plot your results on a graph the result is a curved shape called a **growth curve**. Plot two different growth curves as follows.

8 Soak 40 peas in water for 12 hours, then wrap them in wet paper towelling. Put the wrapped peas in beakers of water in a warm, dark place (see diagram opposite).

9 Every three days remove five seedlings and find their average weight. Then heat them at 100 °C until they are completely dry and find their average weight again. Plot graphs for wet and dry weight changes.

10 Explain why the wet and dry growth curves are different. What conclusions can you draw about the early growth of plants from these results?

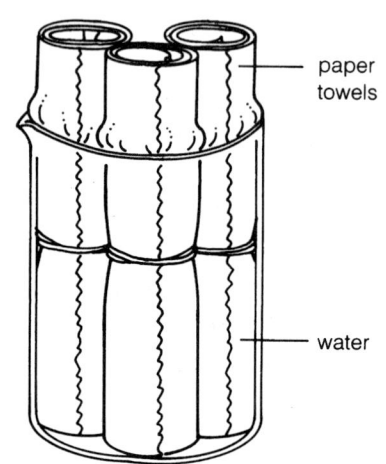

paper towels

water

Your teacher will be looking for: **and especially for:**

- careful use of the apparatus given ☐
- accurate observations and measurements ☐
- good presentation of results including tables and graphs ☐
- sensible conclusions which fit your results ☐
- .... ☐

Bones	BIOLOGY ACTIVITY	A16

You need:

chicken leg bones

safety goggles

retort stands and clamps

dilute hydrochloric acid

masses and mass hangers

Bunsen burner

metre rule

tongs

Investigating bone structure

Bone contains **minerals** (mainly calcium compounds) and **organic fibres**. What is the function of these parts?

1 Obtain two small bones (e.g. chicken leg bones). Remove the minerals from one bone by soaking it in dilute hydrochloric acid for 24 hours. Soak another bone in water for the same time (the control).

2 Pour away the acid, wash the bone and try to bend it. Compare it with the bone soaked in water. What do your observations tell you about the functions of minerals and organic fibres in bones?

3 Remove the organic fibres from another small bone by holding it (with tongs) in a hot Bunsen flame for a few minutes in a fume cupboard.

4 When it is cool, try to crush the burnt end with a pencil or stick., What do your observations tell you about the functions or organic fibres in bones?

Why are backbones arched?

Look at the skeleton of a rabbit or other four-legged animal and note the shape of its backbone. Is there an advantage in having an arched backbone?

5 Support a metre rule (or strip of wood) on chair backs or clamp stands so that each end overlaps the support by 1 cm. Attach weights to its centre until it bends and falls off the supports.

Rabbit skeleton

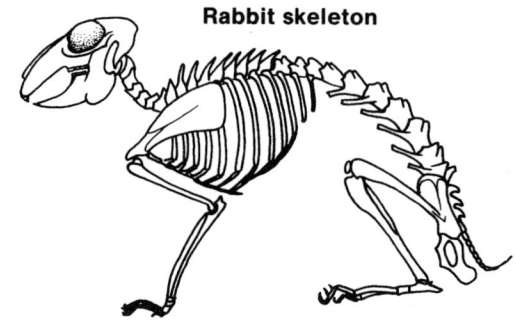

6 Place a metre rule between two supports in such a way that when the supports are moved towards each other the rule arches upwards slightly.

7 Attach weights to its centre as before. What do your results tell you about the advantage of having an arched backbone?

Your teacher will be looking for:	and especially for:
• careful and safe use of the apparatus given	☐
• accurate observation	☐
• good presentation of results	☐
• sensible conclusions	☐
•	☐

HAZARD WARNING

Hydrochloric acid is harmful. Avoid skin contact. WEAR EYE PROTECTION. When breaking glass rods WEAR EYE PROTECTION and use a safety screen.

Natural and manufactured structures

Natural and manufactured structures are designed to cope with the forces they meet: a tree being blown by the wind or a bridge with its load of cars.

Here are some examples of how nature has evolved structures to resist forces. Your task is to spot manufactured structures which are designed in a similar way. Boxes have been left for you to sketch and describe what you find.

1 Spiders have support and protection from the hard skeleton which 'packages' their body.

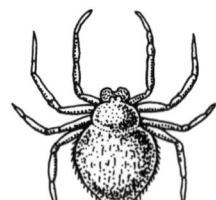

3 The tree trunk gets wider towards the ground as the forces the trunk has to resist get stronger.

2 The large leaf surface area is supported by a network of 'girders', the leaf veins.

4 The twig is very flexible and will therefore bend rather than break in a strong wind.

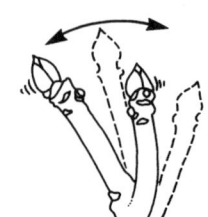

| Strength and muscle fatigue | BIOLOGY ACTIVITY A18 |

You need:

large textbooks or other heavy objects

stop watches or clocks with second hands

A test of strength

1 Work in pairs. One member makes accurate timings, in seconds, while the other performs a task.

2 One member must hold a book or other heavy object in his/her preferred hand (i.e. right hand if right-handed) so that it is at arm's length. The arm must be straight out at the shoulder and must be kept in this position for as long as possible.

3 The time-keeper starts timing as soon as the book is in the required position and stops when the book is lowered or the arm bent. Record how long the weight is held.

4 The pair now swap tasks and repeat this procedure.
The pair swap tasks again but this time the book is held in the *non-preferred hand.*
The pair swap tasks again so that the other member tries the task with his/her non-preferred hand.

5 Results for each pupil should be recorded on the chalkboard, then converted into a table showing how many pupils held the weight, in each hand, for intervals such as 60-69 seconds, 70-79 seconds, etc. up to the longest time recorded.

6 Convert these results into histograms for preferred and non-preferred hands.
Explain any differences between preferred and non-preferred hands.
Explain any differences between the sexes.
What caused fatigue in muscles?

Investigating muscle fatigue

7 Use the results of the previous experiment to find four pupils who are roughly equal in strength.

8 **Pupil one** holds a weight at arm's length, as described in the previous experiment, for as long as possible. He/She rests for 10 seconds and then repeats the task. Continue in this way for five repetitions or until the pupil is too tired to go on. Record, in seconds, the time the weight is held for each repetition.

9 **Pupil two** does the same as pupil one except that a rest of 20 seconds is allowed between repetitions.

10 **Pupil three** does the same as pupil one except that a rest of 30 seconds is allowed between repetitions.

11 **Pupil four** does the same as pupil one except that a rest of 40 seconds is allowed between repetitions.

12 Construct a table showing how long each pupil held the weight during each repetition.

13 **Questions:**
What effect on performance did the different rest periods have?
Try to explain these differences (relate them to the efficiency of the circulatory system, anaerobic respiration and the oxygen debt).

Your teacher will be looking for: and especially for:
• the use of 'fair' tests to measure strength and muscle fatigue ☐
• good presentation of results in tables and as histograms ☐
• sensible conclusions consistent with your results ☐
• .... ☐

HEALTH CHECK

Check with your teacher that you are able to participate in this activity.

Demonstrating respiration

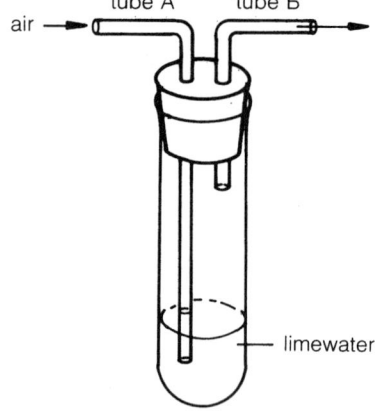

tube A tube B
air →
limewater

You need:

boiling tube and rubber bung with two holes

limewater

glass tubing

four conical flasks and four rubber bungs with two holes each

rubber tubing

Compare breathed and unbreathed air

1 Prepare the apparatus opposite. How long does it take for limewater to turn milky when you blow air gently down tube A? Wash out the boiling tube and refill with fresh limewater. How long does it take for limewater to turn milky when you suck air gently through tube B?

2 **Questions:**

What turns limewater milky?
What do your results tell you about the difference between laboratory air and breathed air?
What exactly do your results prove?

Demonstrate respiration in animals and plants

3 Set up the apparatus below.
What is the purpose of flask 1?

Why is flask 2 necessary?
What does the apparatus demonstrate?

4 Investigate woodlice, maggots, earthworms, etc., in a small specimen chamber. Use a bell jar to investigate bigger animals.

5 **Problem:** How could you use this apparatus to make rough comparisons of respiration rate in, for example, woodlice and earthworms?
What things would you have to keep the same?
How would you set up the apparatus to demonstrate respiration in plants?

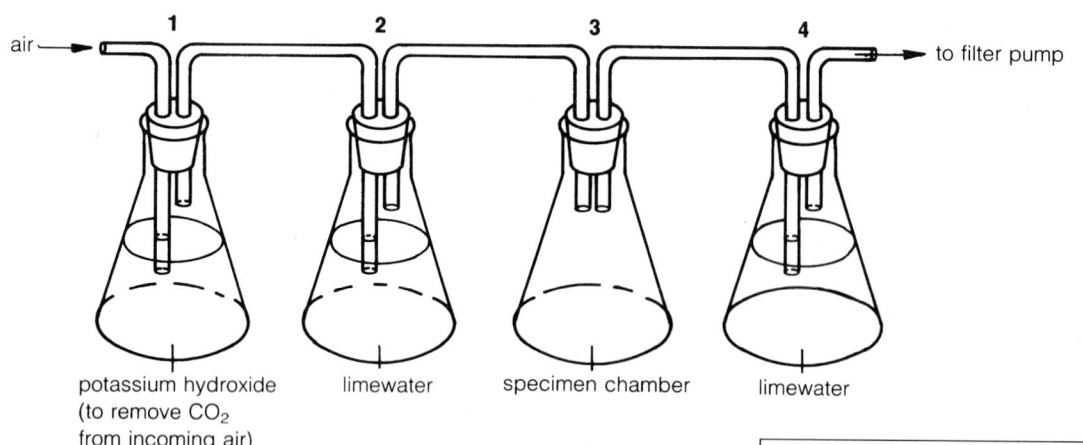

air → 1   2   3   4 → to filter pump

potassium hydroxide
(to remove CO_2
from incoming air)   limewater   specimen chamber   limewater

Your teacher will be looking for: and especially for:

- accurate observation
- good presentation of results
- care with live specimens
- sensible conclusions which match your results
- ....

HAZARD WARNING

Soda lime is CORROSIVE and can cause severe burns; also dangerous to eyes and skin. AVOID SKIN CONTACT. AVOID CONTACT WITH WATER. WEAR EYE PROTECTION.

Measuring respiration	BIOLOGY *ACTIVITY*	A20

You need:

specimen tubes and bungs
  with two holes

soda lime

50 cm³ beakers

capillary tubing

perforated zinc

ink (coloured water)

white card

sticky tape

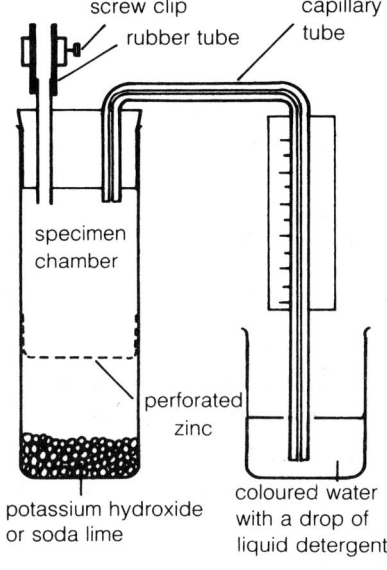

screw clip

rubber tube

capillary tube

specimen chamber

perforated zinc

potassium hydroxide or soda lime

coloured water with a drop of liquid detergent

1 Prepare the apparatus opposite. It is a simple **respirometer**; that is, it can be used to measure the rate of respiration. Predict what will happen to coloured liquid in the capillary tube if you put a few maggots or other small creatures in the specimen chamber and close the screw clip.
What is the reasoning behind your prediction?

2 Carry out this experiment and check your prediction. Aerobic organisms take in oxygen and produce carbon dioxide at about the same rate. Carbon dioxide is absorbed by the soda lime. Use these facts to explain your result.

Find out: How to measure respiration rate in a variety of small animals

What exactly does this apparatus measure?

Design an experiment to measure and compare respiration rate in small creatures such as woodlice, beetles, spiders, earthworms, maggots, etc.

Start by thinking about the following:
What conditions must be kept the same for each creature so that results can be compared?
How could you improve your results if you knew the diameter (bore) of the capillary tube?
Why is the rubber tube and screw clip essential, and when should the clip be opened and closed?

Your teacher will be looking for:	**and especially for:**
• careful use of the apparatus given	☐
• suitable treatment of living specimens	☐
• accurate observation and measurement	☐
• good presentation of results	☐
• sensible conclusions	☐
•	☐

HAZARD WARNING

Soda lime is CORROSIVE and can cause severe burns; also dangerous to eyes and skin. AVOID SKIN CONTACT. AVOID CONTACT WITH WATER. WEAR EYE PROTECTION.

A fishy problem

Some solids and liquids will dissolve in water. The graph opposite shows how the amount of salt which will dissolve in water changes as the temperature increases. The dotted line shows how the amount of oxygen which will dissolve changes as the temperature increases.

1 What do you notice?

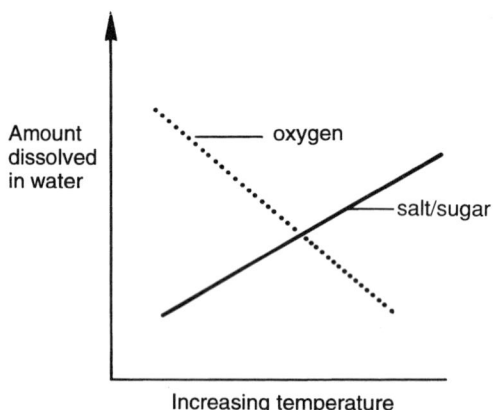

2 Would a fish need to have more efficient gills in warm or cold water? Explain your answer.

A fish gets the oxygen it needs for respiration from the water.

3 Fish are cold-blooded. This means that they become more active as they warm up. What substances will their cells need more of as the fish move about more? Why?

4 What will happen to the amount of oxygen available to the fish as the surrounding water warms up?

5 What problem do your last two answers highlight about fish living in warmer water?

6 Explain why a Minnow (a small, British freshwater fish) is unlikely to survive in a tropical river.

Further work

a) The Dead Sea is very warm and is a saturated salt solution. If you put a cup of clear Dead Sea water into a refrigerator (not a freezer!), what would you expect to see in the morning? Explain your answer.

b) If you put sugar in your tea, would it dissolve more quickly if you added it before or after the milk? Explain your answer.

Anaerobic respiration

You need:

test tubes	yeast in cold, boiled water
thermometer	glucose in cold, boiled water
balloons	fresh peas in cold, boiled water
measuring cylinder	boiled peas in cold, boiled water
limewater	containing bactericide (eg. sodium
liquid paraffin	chlorate(I) 10%)
Thermos flasks	

Anaerobic respiration (fermentation) in yeast

1 Label three test tubes A, B, and C.
   Place 20 cm³ of yeast suspended in cold, boiled water into tube A. (Boiling removes oxygen from water.) Add a few drops of liquid paraffin (to cover the surface of the yeast suspension).
   Place 20 cm³ of glucose dissolved in cold, boiled water into tube B. Add liquid paraffin.
   Place 10 cm³ of yeast suspended in cold, boiled water into tube C, then add 10 cm³ of glucose dissolved in cold, boiled water. Mix the two together. Add a few drops of liquid paraffin.

2 Place a balloon firmly over the neck of each tube (tie with cotton if necessary). Make sure the balloon is deflated. Put the tubes in a warm place for 24 hours.

3 Record what happens to the balloons.
   Explain your observations.
   Why was cold, boiled water used in this experiment?
   Why was liquid paraffin added to each tube?
   What gas entered a balloon?

balloon

liquid paraffin

yeast/glucose mixture

Anaerobic respiration in peas

4 Fit two Thermos flasks with bungs through which a thermometer has been passed. Label the flasks A and B.

5 Fill flask A with boiled peas in cold, boiled water containing bactericide.
   Fill flask B with fresh peas in cold, boiled water.
   Note the temperature of each flask.

6 Place the bung in each flask, so that air cannot enter and leave them for a week.
   Note any temperature changes daily.

7 Explain any changes which occur in the temperature, and in the peas.
   Why was one set of peas boiled?
   Why were both sets of peas in boiled, cold water.
   Why was bactericide added to flask A?

Your teacher will be looking for:	and especially for:
• careful and skilful use of the apparatus given	☐
• good observation and presentation of results	☐
• sensible conclusions which match your results	☐
•	☐

HAZARD WARNING

Sodium chlorate(I) solutions is corrosive. AVOID SKIN CONTACT. WEAR EYE PROTECTION.

Lung volume

You need:

bell jar with 500 cm³ graduations down one side

50% alcohol

rubber tube

deep sink

Breathe in as much air as you can, then breathe out as much air as you can. The volume of air which you breathed out is called the **vital capacity** of your lungs.

After you have forced as much air as possible out of your lungs there is still about 1500 cm³ left behind. This is called the **residual volume** of your lungs.

The **total volume** of your lungs is your vital capacity *plus* your residual volume.

1 Mark out the side of a bell jar in 500 cm³ units as follows:
  Place a bung firmly in a bell jar and turn it upside down in a sink (bung downwards). Pour 500 cm³ of water into the bell jar and mark the water level with a chinagraph pencil or indelible marker.
  Repeat until the whole of one side is marked out.

2 Fill a deep sink with water.
  Place the bell jar on its side in the water so that it is completely full.
  Without letting in any air, turn the bell jar upright in the sink. Run water out of the sink until it is about half full.

3 Work in pairs.
  One pupil tilts the bell jar enough to let in a rubber tube (see diagram) and then holds the tube in place.
  The other pupil breathes in as much air as possible and then blows out as much air as possible through the rubber tube into the bell jar.

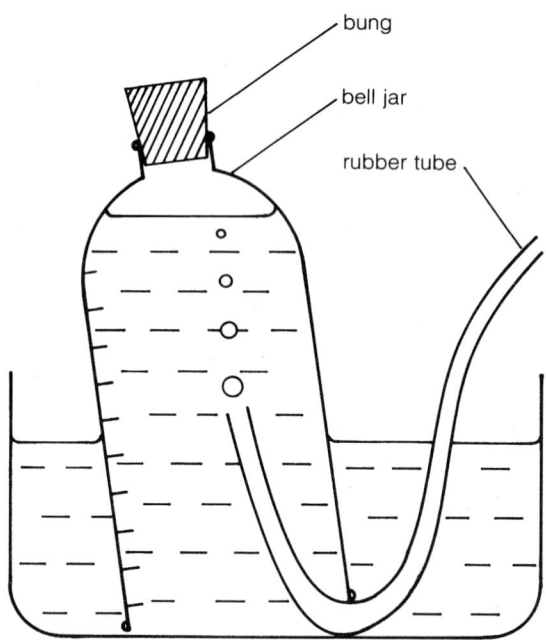

Raise the bell jar so that the level of water inside is level with water in the sink, and read off and record the vital capacity of your lungs.
Sterilize the end of the rubber tube in alcohol, then wash it in water.
Swap tasks and repeat this procedure.
Prepare a class results table on the blackboard.

4 What is the total volume of your lungs? What is the average vital capacity of your class?
  What is the smallest and the largest vital capacity in your class?

Your teacher will be looking for:	**and especially for:**

- accurate measurement and recording of results ☐
- good presentation of class results and accurate calculation ☐
- .... ☐

HAZARD WARNING

Alcohol is highly flammable. KEEP AWAY from naked flame. WEAR EYE PROTECTION.

| Pulse and breathing rates | | BIOLOGY *ACTIVITY* | A24 |

Pulse and breathing rates

You need:

clock or watch with a second hand   bench or step about 30 cm high

A standardised fitness test

(Check with your teacher that you should do these exercises *before* beginning this activity.)

1 Form groups of three. You will take it in turns to do a standardised exercise. (The two members of the group who are not exercising keep time and measure breathing and pulse rates.)

2 Before you start, measure and record your resting pulse rate and your resting breathing rate using the methods below:

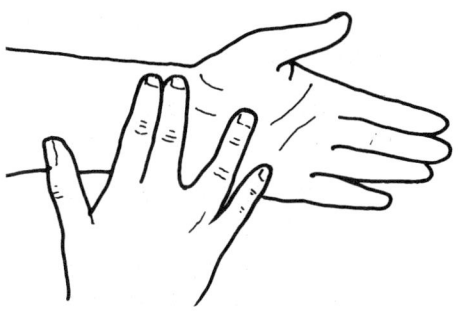

> **Pulse rate:** Place your fingers as shown in the diagram. Move them until you can feel a pulse. Count the pulses in 30 seconds, then multiply by two to get the pulse rate per minute.
>
> **Breathing rate:** Count the number of breaths taken in 30 seconds and multiply by two.

3 One member performs the following exercise while another calls out the time at one minute intervals.

4 Face the bench. Step onto it with one foot, then step up with the other foot. Step down with the first foot, then down with the second foot. Practise this so that you can step steadily at about 25 steps per minute.

5 When you and the timer are ready, start stepping and keep going at a steady speed for three minutes.

6 *Immediately* afterwards, one person measures the pulse rate of the 'exerciser'. At the same time, the other team member measures the exerciser's breathing rate. Keep a record of your results.

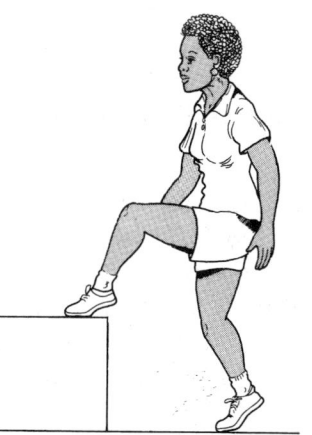

7 Continue monitoring the pulse rate and breathing rate at one minute intervals until they return to normal.

8 Swap places and repeat steps 2 to 6 until you have all done the test.

HEALTH CHECK

Check with your teacher that you are able to participate in this activity.

Questions:

- What are the fastest, slowest, and average *resting* pulse rates for the class?
- What are the fastest, slowest, and average *resting* breathing rates for the class?
- What are the fastest, slowest, and average pulse rates *after exercise* for the class?
- What are the fastest, slowest, and average breathing rates *after exercise* for the class?
- Is the fittest person the one with the lowest or the highest pulse rate?
- How does fitness affect the time taken for the pulse to return to normal?

The functions of perspiration

You need:

thermometers   cotton wool   small beakers of water and alcohol

1 Wave your hand backwards and forwards in front of you.
   Does it feel warmer or cooler?

2 Swab a little water onto the back of one hand with cotton wool
   and wave it backwards and forwards again.
   What does the wet part feel like compared with the dry parts?
   Note any feelings to do with temperature sense.

3 Dry your hand, then swab a little alcohol onto it with cotton
   wool. Wave it backwards and forwards again.
   What does the alcohol-treated part feel like compared with the
   dry parts?
   Note any feelings to do with temperature sense.
   Does the alcohol give you a sensation which is different from
   water?

4 What conclusion can you draw at this stage of the experiment?

5 Obtain three thermometers. Record the temperature of each.
   Wrap the bulbs of all three thermometers in a thin layer of
   cotton wool. Tie it in place with cotton.
   Leave the thermometers to acclimatize for five minutes, then
   record the temperature reading of each.

6 Take one thermometer (with dry cotton wool) and wave it for
   one minute, then record the temperature reading.
   Take another thermometer, dip it into water, wave it for one
   minute, then record the temperature reading.
   Take a third thermometer, dip it in alcohol, wave it for one
   minute, then record the temperature reading.

7 **Questions:**
   How do the results with a thermometer dipped in water and
   alcohol compare with those from the thermometer covered
   with dry cotton wool?
   How do the results of the thermometer experiment help
   explain the results from treating skin with water and alcohol?
   Alcohol evaporates more quickly than water. Use this
   information to explain results from both experiments.
   What do these results tell you about the function of
   perspiration?

? A chemist has
invented a new,
harmless liquid that
evaporates quickly.
She wants to sell it in
swabs which people
can use to wipe
themselves when they
are too hot.
Design an experiment
to compare the
effectiveness of this
invention with a wet
cloth.

Your teacher will be looking for:	**and especially for:**
• careful and safe use of the apparatus given	☐
• accurate observation	☐
• good presentation of results	☐
• sensible conclusions	☐
•	☐

HAZARD WARNING

Alcohol is highly
flammable. KEEP AWAY
from naked flame. WEAR
EYE PROTECTION.

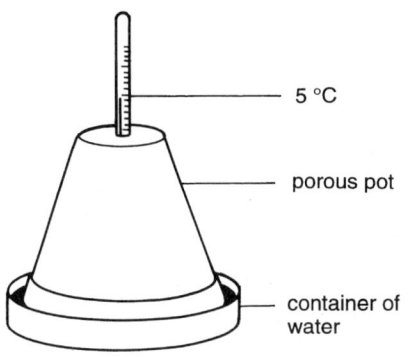

A bottle of milk can be kept cool using this simple method:
The bottle of milk is put into the shallow container of water and
is covered with the porous pot. The clay pot soaks up the water.
Even in a warm place, the inside of the milk cooler stays at about
5 °C.

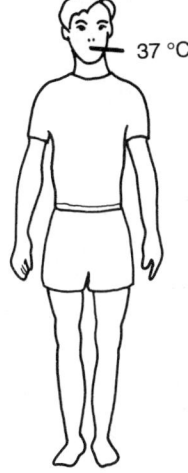

Sweating is one of the control systems which helps humans to
keep cool. Water from the body forms sweat on the surface of the
skin. This evaporates and cools us down.

1 Why does the milk cooler stay at 5 °C? (Use the words 'water',
   'evaporate', 'energy' and 'temperature' in your answer.)

2 How are the milk cooler and the body temperature control
   system similar?

3 As humans sweat, why don't they cool down to 5 °C like the
   milk cooler?

4 When people are ill, it sometimes helps to put a damp flannel
   on their forehead. How does this help to reduce the body
   temperature?

? Cheddar cheese used to be matured in caves where the
temperature is a constant 10 °C. This is the perfect
temperature for maturing cheese. Design an experiment to
find out whether the 'milk cooler' could be adapted to give
a temperature of 10 °C.

Write down the things which should be investigated and
the steps that you would take.

Sensitivity to temperature

You need:

250 cm³ beakers   ice-cold water

thermometers   hot water

bulb pipettes   sticky labels

Are some areas of your hand more sensitive than others to differences in temperature?

1　Label three beakers 25 °C, 35 °C, and 45 °C. Use hot and ice-cold water to fill each beaker with water at these temperatures and keep them at this level during the experiment. Put a bulb pipette in each beaker.

2　The idea is to test the skin of one hand with water at different temperatures.
　Work in pairs. One member sits facing away from the water samples, with a hand on the bench, while the other performs the tests. Test:
　　fingertips
　　palm
　　back of the fingers
　　back of the hand

3　Use a bulb pipette to place one drop of water carefully onto the area being tested. Your partner must say if it is hot, or warm, or cold. Note the reply.
　Test each area a total of 12 times, made up of four tests at each temperature. Choose different temperatures at random so that your partner does not know which to expect.

4　Design a suitable results table with correct and incorrect responses for each area recorded as ticks and crosses.

5　List the areas you tested from the most to the least sensitive. Explain your results.
　Why is the temperature sense useful?

Investigating sensory adaptation

6　Label three 250 cm³ beakers 'cold', 'warm' and 'hot'.
　Fill the first with ice-cold water, the second with water at room temperature, and the third with hot (*not boiling*) water.

7　Place the beakers in a row with the water at room temperature in the middle.
　Place the fingers of one hand in the hot water, and the fingers of the other hand in the cold water.
　After one minute, place the fingers of both hands into the water at room temperature.

8　**Questions:**
　Does the water feel different to each hand? Can you explain this result?
　A traveller from the arctic and another from the tropics arrived in London on a warm spring day. The traveller from the arctic thought that London's weather was warm, but the traveller from the tropics thought it was cold. Use your results to explain their feelings.
　What does **sensory adaptation** mean?

Your teacher will be looking for: **and especially for:**
- careful use of the apparatus given ☐
- accurate observations ☐
- good presentation of results ☐
- sensible conclusions ☐
- .... ☐

Food tests I

You need:

iodine	liquid egg albumen
Benedict's reagent	bread
Biuret reagent	potatoes
ethanol	cheese
test tubes and racks	cooking oil
spotting tiles	glucose
Bunsens, tripods and gauzes	starch powder
250 cm³ beakers	milk powder
goggles	peanuts
glass rods	suet
spatulas	peas and beans
bulb pipettes	carrots
sodium hydrogen carbonate	grapes

Begin by performing the following tests on known foods to observe a **positive result**. It is recommended that you then repeat each test with sodium bicarbonate to observe a **negative result**. These observations will be helpful when you go on to test foods of unknown composition.

1   **Test for starch**
    Place a little starch powder in a depression on a spotting tile. Add a few drops of iodine.
    Positive result: blue/black colour
    Negative result: brown colour

2   **Test for glucose**
    Place equal quantities of a strong glucose solution and Benedict's solution in a test tube (about 2 cm³ of each). Lower the test tube into a beaker of boiling water, wait until the test tube contents boil and leave it for two minutes.
    Strong positive result: brick red precipitate
    Medium positive result: yellow orange precipitate
    Weak result: green colour
    Negative result: blue colour
    Before testing a solid food it must be crushed in warm water to extract any glucose which may be present.

HAZARD WARNING

Iodine and Benedict's are harmful to skin and eyes. AVOID SKIN CONTACT. WEAR EYE PROTECTION.

---

3 Test for proteins

Dissolve a little milk powder in water in a test tube. Add a few drops of Biuret reagent.
Positive result: purple colour
Negative result: blue colour
Note that this is a test for soluble proteins. Before testing a solid food it must be crushed in warm water to dissolve any proteins which may be present.

4 Test for oil and fat

Place about 1 cm³ of ethanol in a test tube. Add a few drops of oil and mix by shaking. Add an equal amount of water and shake again.
Positive result: a cloudy emulsion forms
Negative result: liquid remains clear
Food containing solid fats are tested by crushing them in ethanol to obtain an alcoholic solution. This is filtered and added to water.

More things to do

Use these tests to analyse the range of foods provided.
Divide each food into four samples and perform one test on each.
Remember to crush solid samples in warm water to extract glucose and protein, and alcohol to extract fats and oils.
Make sure you know the difference between positive and negative results.
Design a results table to show positive and negative results for each test.
List the types of food found in each sample.

From your tests list the types of food present in bread, milk, boiled potato and cheese.
Would eating these foods give you a balanced diet (i.e. do they contain sufficient carbohydrate, protein and fat for health)?
What would be the result of basing your diet on these foods alone?

Your teacher will be looking for:	**and especially for:**
• careful and safe use of the apparatus given	☐
• good observation	☐
• good presentation of results	☐
•	☐

HAZARD WARNING

Ethanol is highly flammable. KEEP AWAY from naked flame. Biuret is harmful to skin and eyes. AVOID SKIN CONTACT. WEAR EYE PROTECTION.

Measuring the energy values of foods

You need:

Bunsen burner	thermometers
wood splints	safety goggles, screen
stands and clamps	measuring cylinder
boiling tubes	foods: peanuts,
mounted needles	sunflower seeds, bread

1 Put 20 cm³ of water into a boiling tube. Fix the tube in a clamp so that it is held at an angle of 45° (see diagram).

2 Weigh a peanut very carefully, in grams (if possible to two decimal places), and note the result.

3 Fix the peanut onto a mounted needle, taking care that no bits drop off.

4 Measure the temperature of the water in the boiling tube and note the result.

5 Ignite the peanut in a Bunsen flame. *Quickly* place the burning peanut under the boiling tube. The idea is to use as much heat as possible from the burning nut to heat the water in the tube.
If the nut goes out, relight it quickly and put it back under the tube.
When the peanut has completely burnt, measure the temperature of water in the boiling tube again and note the result.

6 Before you can go any further you must know:
   • the mass (weight) of water in the boiling tube (1 cm³ of water weighs 1 g)
   • the rise in temperature of water in the boiling tube
   • the mass (weight) of the peanut

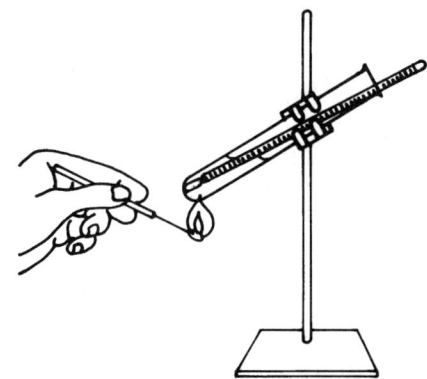

7 It takes 4.2 joules of energy to raise the temperature of 1 g of water by 1 °C, therefore you can calculate the energy given off by 1 g of peanut as follows:

$$\frac{\textbf{mass of water (in grams)} \times \textbf{rise in temperature} \times \textbf{4.2}}{\textbf{mass of the peanut}}$$

8 Your result will be much lower than the actual energy value of 1 g of peanut. Give as many reasons as you can why this is so. Despite this fact, if you use this method to find out the energy value of other foods, your results can still be compared. Why is this so?

9 Use this method to find out the energy values (in joule per gram) of the foods provided. Produce a results table and comment on your findings.

10 Design an improved method which will give a more accurate result. (Hint: is there any way of reducing heat loss to the air?)

Your teacher will be looking for: **and especially for:**
* careful use of the apparatus given ☐
* accurate measurements of volume of water, mass of food, temperature ☐
* accurate recording of results and successful calculations of energies ☐
* critical evaluation of the experiment and sensible suggestions for improving it ☐
* .... ☐

HAZARD WARNING

Wear EYE PROTECTION when burning foods. Use a safety screen.

You need:

visking tubing	starch solution
boiling tubes	amylase solution
test tubes	Benedict's reagent
500 cm³/250 cm³ beakers	iodine solution
thermometers	Bunsens, tripods and gauzes
wood splints	spotting tiles
sticky labels	cotton thread

1 Half fill a 500 cm³ beaker with water and heat it to 37 °C. Keep it at this temperature by adding hot and cold water.

2 Label three boiling tubes A, B, and C, fill them with water at 37 °C and put them in the beaker of water at this temperature.

3 Open three 15 cm lengths of visking tubing by rubbing them between your fingers under a running tap and tie one end of each securely with cotton.
Fill one length of visking tubing with starch solution, tie the open end with cotton, rinse it under a tap and place it in tube A.
Fill another length of visking tubing with amylase solution, tie the open end with cotton, rinse it under a tap and place it in tube B.
Fill the last length of visking tubing with equal amounts of well-mixed starch and amylase solutions, tie one end with cotton, rinse it under a tap and put it in tube C.
Make sure the visking tubing in each boiling tube is completely covered with water. Place all three in the 500 cm³ beaker of water maintained at 37 °C.

4 Now test samples from the remaining solutions. Test the starch solution with iodine and Benedict's reagent (page 38).
Test the amylase solution with iodine and Benedict's reagent.

5 After 10 minutes obtain water from around the visking tube in tubes A, B and C. Test each sample with iodine and Benedict's reagent.
Repeat after a further 10 minutes.
Devise a results table.

6 **Questions:**
What are iodine and Benedict's reagent used to test for?
Why were the starch and amylase solutions tested with iodine and Benedict's reagent?
What do your results tell you about amylase, about visking tubing, and about the substance which passed through the visking tubing?
What part of the body does visking tubing represent?
Why are digestive enzymes necessary?

Further work

Design experiments to investigate the effects of temperature and pH on amylase.

Does it work best at body temperature and will it work above or below this temperature?

Does it work best in acid, neutral, or alkaline conditions?

Your teacher will be looking for:	and especially for:	
• careful and skilful use of the apparatus given		☐
• accurate observation and recording of results		☐
• good presentation of results in tables		☐
• sensible conclusions		☐
•		☐

HAZARD WARNING

Amylase is an irritant.
AVOID SKIN CONTACT.
Iodine and Benedict's are harmful to skin and eyes.
AVOID SKIN CONTACT.
WEAR EYE PROTECTION.

Investigating the sense of taste	**BIOLOGY** **ACTIVITY**	**A32**

You need:

50 cm³ beaker labelled 'sweet', containing sugar solution

50 cm³ beaker labelled 'salt', containing salt solution

50 cm³ beaker labelled 'sour', containing citric acid solution

50 cm³ beaker labelled 'bitter', containing strong black coffee

Your tongue has taste receptors, called **taste buds**. There are four kinds of taste bud: those which are sensitive to sweet, salt, sour and bitter tastes. This experiment shows you how to find out if the different kinds of taste bud are evenly spread over the tongue, or if certain areas have only one type.

This experiment should ideally be done in a classroom.

1 Copy the diagram of a tongue from the illustration opposite.

2 Work in pairs. One member of the pair, the subject, sits with eyes closed and his or her tongue sticking out.
The other member, the experimenter, uses a sterile plastic bulb pipette to place *one drop* of liquid from a beaker onto the subject's tongue. The subject *must not know* which taste is being tested.
The subject says which taste has been used.
The experimenter notes the reply on a results table which should show:
area of tongue tested
taste tested
correct/incorrect reply
ease of reply (i.e. if the subject easily
    detected the taste or had difficulty
    and had to guess).

3 The mouth should be rinsed between tests.
Each taste solution should be tested ten times – twice in every area shown on the diagram.

4 **Questions:**
How do your results compare with the rest of the class?
Do the areas marked on the diagram of the tongue correspond with one or more tastes?
If there are only four different 'tastes', how are all the hundreds of different 'flavours' of food and drink produced?
Why is taste necessary?

Map of the tongue's taste areas

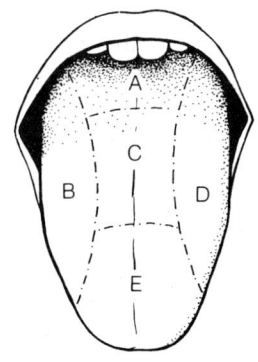

Your teacher will be looking for:	**and especially for:**
• careful use of the apparatus given	☐
• good presentation of results in a table	☐
• sensible conclusions consistent with your results	☐
•	☐

Balance and hearing

You need:

clear plastic tubing
  long enough to
  make a ring 5 cm
  in diameter

whistle

blindfold

Investigating the organs of balance

1 Obtain a piece of wood or plastic which can be used to join the ends of the tube to make a watertight ring. Fill the tube with water but allow a small bubble of air to remain, then connect the tube ends.

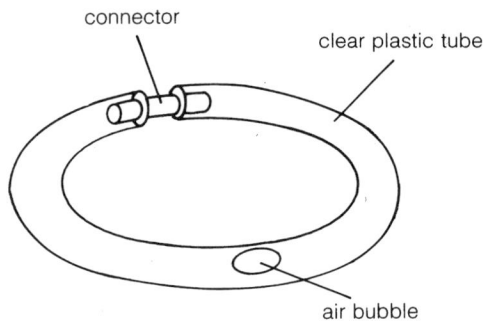

connector

clear plastic tube

air bubble

2 **Questions:**
What part of the inner ear does the tube represent?
Hold the tube horizontally, then tilt it from side to side. What happens to the bubble?
If the tube was lined with sensitive hairs, what would happen to them as the tube is tilted in this way?
How is this demonstration similar to what happens in your inner ear when you tilt your head from side to side?
Spin (**rotate**) the tube horizontally on a table top, then stop it suddenly. What happens to the bubble?
What does this tell you about why people feel dizzy if they spin round and round and stop suddenly?

Why do we have two ears?

3 **Method one:** A blind-folded student sits in the centre of a classroom while another student, without shoes, moves *quietly* about the room making a noise, such as a clap, from different positions.
The blind-folded student tries to point in the direction of the claps.
Repeat the experiment while the blind-folded student covers one ear.

4 **Questions:**
Using two ears, is it more difficult to judge the direction of some claps than others? If so, which direction do these claps come from?
Look at the shape of the ears. Does this help explain your results?
Are your results different when one ear is covered? If so, what does this tell you about why we have two ears?

5 **Method two:** A group of students line up on a playing field facing a teacher about 50 paces away.
- The students are blind-folded and try to find the teacher from memory.
- Blind-folded students try to find the teacher while he/she blows a whistle at one-second intervals.
- Blind-folded students with *one ear covered*, try to find the teacher while he/she blows a whistle.

6 **Questions:**
Compare your three results. What do they tell you about why we need two ears?
Using both ears, how do you know which direction a sound is coming from?
How will covering one ear upset direction-finding?

| Eyes and vision I | BIOLOGY ACTIVITY | A34 |

Eyes and vision I

You need:

sheep eyes from butcher
  or slaughter house

dissecting trays

scalpels

dissecting scissors

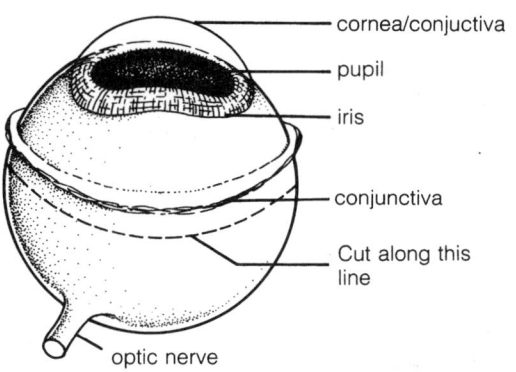

cornea/conjuctiva
pupil
iris
conjunctiva
Cut along this line
optic nerve

Eye dissecting

1 Carefully remove any fat surrounding the eye.
   What are the muscles for, which are attached to the outside of the eyeball?
   Look for the **optic nerve** – it should look like a thick, yellow strand.

2 Using a sharp scalpel and/or scissors, cut through the tough, outer **sclerotic layer** somewhere on the line marked on the diagram opposite.
   Taking care not to squeeze the eye, continue cutting along this line until the eye can be separated into front and back halves.
   Lay the two portions in a dissecting tray of water so that the insides of front and back can be examined.

3 **Questions:**
   What is the jelly-like substance in the back of the eye?
   What is the colour of the inside surface of the eye? Why is it this colour?
   Find the point inside the back of the eye where the optic nerve leaves the eye. What is this point called?
   What are the black fibres which radiate outwards from the lens?
   Remove the lens and observe its shape. What happens to its shape when the eye is focused on near and distant objects?
   Look for the **iris** and **pupil**. What is the iris made of and what is its function?

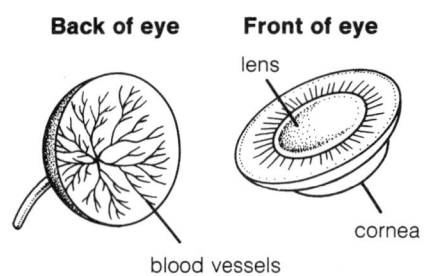

Back of eye **Front of eye**

lens

cornea

blood vessels

Find your blind spot

4 Hold this book at arm's length. Close your left eye and stare at the cross below with your right eye. Note that, without looking directly at it, the black circle can still be seen. Bring the book slowly towards your face *(don't look away from the cross)*. At a certain point the circle disappears. This happens when its image falls on the blind spot.

5 Why don't your blind spots stop you seeing properly?

+   ●

Your teacher will be looking for:	**and especially for:**
• careful and skilful use of the apparatus given	☐
• accurate observation	☐
• good presentation of results	☐
•	☐

HAZARD WARNING

Scalpels are sharp, handle with care.

Eyes and vision II

You need:

two pencils　　　　flat desk or table top

large coins

Distance judgement

This experiment shows you how to compare one eye with two eyes when judging distances. It should help you understand why we have two eyes.

1　Arrange two pencils on a desk top in positions A and B, as shown on the diagram opposite.

2　Sit so that your eyes are level with the surface of the desk (it is very important that you do not look down on the desk).

3　The aim is to move the pencils until their points are exactly opposite but *not touching* (i.e. to positions A$_1$ and B$_1$). Do this in three different ways:
   Method one: Close one eye and ask a partner to move the pencils by *following your instructions only* (they must not try to correct your mistakes). Return the pencils to positions A and B.
   Method two: Close one eye but this time use *one hand* to move the pencils.
   Method three: Try the experiment again using one hand but *both eyes*.

4　**Questions:**
   Is there any difference between using one eye and two? If so, try to explain this difference. What does this tell you about why we have two eyes?
   Is it easier or more difficult to use one hand or a partner to get the pencils opposite? If so, try to explain any difference.

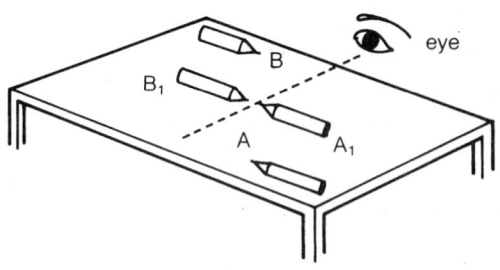

Three-dimensional vision

3-D vision allows you to see rounded, solid objects rather than a flat picture of your surroundings, like a photograph. This experiment helps you understand how your eyes and brain produce a 3-D vision.

5　Hold a large coin with its edge towards you, about 30 cm in front of your eyes.

6　Look at the coin first with your left eye closed and then with your right eye closed.

7　**Questions:**
   What is the difference between the two views of the coin?
   Why does your brain need these two views to produce a 3-D vision?
   Try to think of reasons why, despite these results, your vision does not become completely flat and two-dimensional when you close one eye. (Do we need two eyes after all?)

Your teacher will be looking for:　　　**and especially for:**

- accurate observation ☐
- good presentation of results and sensible conclusions ☐
- .... ☐

Measuring the speed of reflexes

You need:

metric rulers

Measure the speed of your reflexes

1 Work in pairs. One student holds a ruler between thumb and forefinger so that the ruler hangs with its zero mark at the bottom. The other waits with thumb and forefinger of one hand about 2 cm apart and level with the zero mark of the ruler.

2 The student holding the ruler says 'ready', then drops the ruler within five seconds without further warning.
   The other student must catch the ruler between thumb and forefinger.
   Note the number of centimetres the ruler has dropped by looking at the position of the thumb and forefinger on the ruler.

3 Calculate the average distance over at least ten ruler drops. Use the graph opposite to convert this distance into response time, in seconds.
   Draw a graph showing the range of results for the whole class.

4 **Questions:**
   Name all the parts of the nervous system which impulses travel through as you respond to the ruler dropping.
   Your result is the time it takes for impulses to travel from your eyes to your hand. Measure this distance and use it to calculate the speed of nerve impulses, in metres per second.

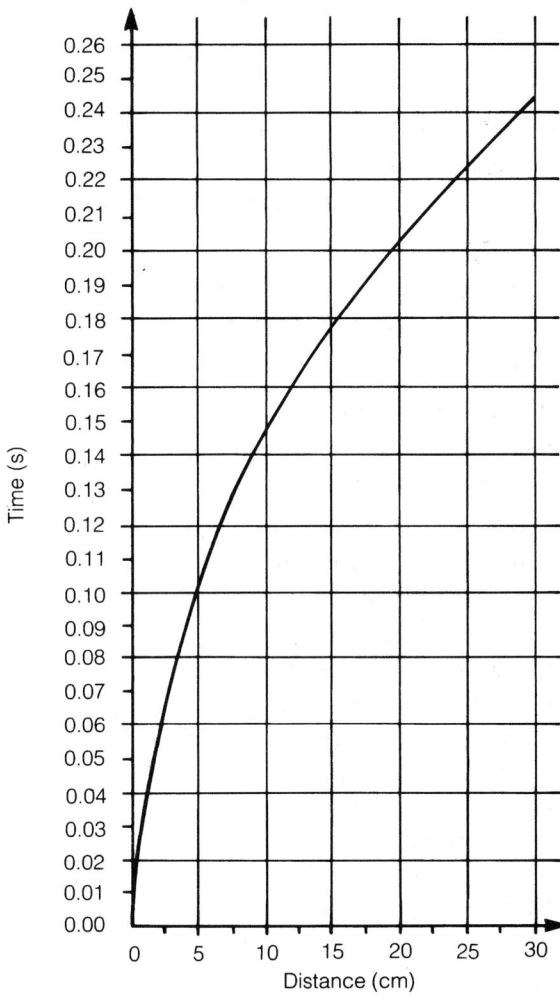

Your teacher will be looking for: and especially for:

- careful collection of data ☐
- good presentation of results ☐
- accurate use of the conversion graph ☐
- successful calculation of the approximate speed of nerve impulses ☐
- .... ☐

Further work

The 'detector' in the experiment above was the eye. Design an experiment to measure the speed of reaction when the signal is a sound detected by the ear.

Pollution

You need:

sticky tape	white card
filter paper	white paper tissues
graph paper	funnels and beakers
pH indicator papers	plankton nets
white pie dishes	note books

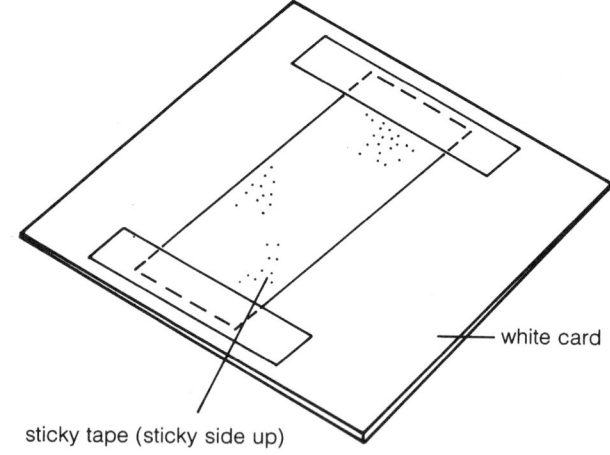

white card

sticky tape (sticky side up)

Air pollution – sticky tape method

The amount of dust, soot and other particles in the air can be investigated by exposing sticky tape to the atmosphere.

1 Fasten a length of transparent sticky tape, sticky side up, to a piece of white card. Leave it outside in dry weather for a day, then examine it under a microscope. Can you see the difference between dust, soot, fibres and other types of dirt?

2 Calculate the area of the field of view: measure its diameter by focusing on the edge of a millimetre rule and divide this by two to find the radius. The formula $A = \pi r^2$ gives the area. How many particles are visible in this area?

Investigating air pollution

How many particles would there be in a square millimetre, and a square centimetre? Use this method to compare air pollution in a number of places from a city centre to open countryside. Can you pinpoint areas of heavy pollution where you live?
Investigate air pollution at various distances from the edge of a busy road.

Air pollution – leaf-wiping method

3 Find an evergreen shrub (privet, holly, etc.), dampen a white paper tissue and use it to wipe a leaf.
Try new leaves and old leaves. Can you see any difference in the black marks on the paper?
Why use evergreen shrubs?

Find out: About dirt on leaves

Measure the amount of dirt on a shrub's leaves in grams per square metre.
Use the leaf-wiping method to solve this problem, then compare the amount of dirt on shrubs from a city centre outwards.
Start by thinking about the following:
Decide how many leaves you should wipe on each shrub.
How will you weigh the amount of dirt on its leaves?
How could you use graph paper to estimate the area of a leaf?
What would be the best time of year to do this investigation?
Why is it only a very rough estimate of air pollution?

Your teacher will be looking for:	and especially for:
• careful use of the apparatus given	☐
• accurate observation	☐
• good presentation of results (including calculations)	☐
• sensible conclusions	☐
•	☐

Acid rain

1 Collect rain as it falls in clean glass jars.

2 Test the water with universal pH test papers to measure its acidity. (Note that pure rain has a pH of 5.7 because it absorbs carbon dioxide from the air to become carbonic acid.)

With dry fingers, tear out one leaf from a book of test papers. Dip the paper into the water and quickly put it on a white tile. Compare its colour with the chart provided with the test papers.

Water pollution

The cleanliness of a stream or pond can often be judged by studying 'indicator animals' known to tolerate different amounts of water pollution. Use the chart below and the drawings to grade streams and ponds into categories from A (clean) to E (completely dead).

Questions:

3 Does city centre rain have a different pH from countryside rain?
Is there a difference in pH between day-time and night-time rain?
Are streams and ponds in your area acidified? (Fresh water should have a pH of about 6.5.)

Freshwater shrimp

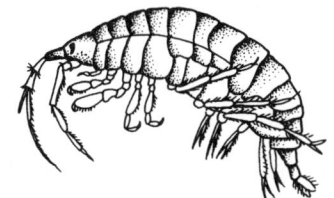

Ostracod

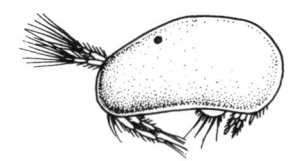

indicator animals	pollution category	pollution status
no apparent life	E	heavy polluted
sludge worms rat-tailed maggots	D	serious pollution
animals from D plus: ostracods blood worms	C	fairly serious pollution
animals from C and D plus: caddisfly larvae freshwater shrimps	B	some pollution
animals from B to D plus: stonefly nymphs mayfly nymphs	A	clean

Bloodworm (midge larva)

Sludge worm

Rat-tailed maggot (drone fly larva)

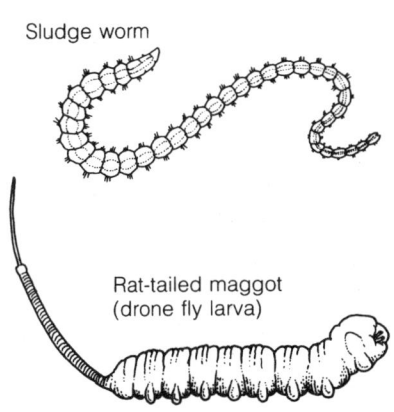

Mayfly larva

Stonefly larva

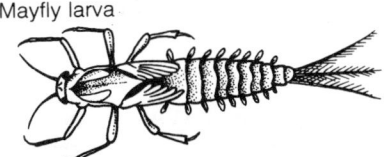
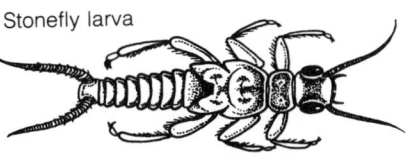

Caddisfly larva

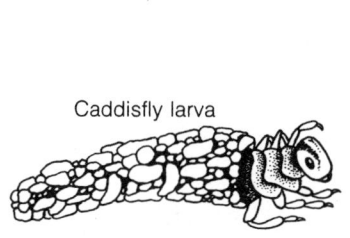

Your teacher will be looking for:	and especially for:
• accurate observations and measurements	☐
• good presentation of results	☐
• sensible conclusions	☐
•	☐

Comparing soil samples I	BIOLOGY ACTIVITY	A39

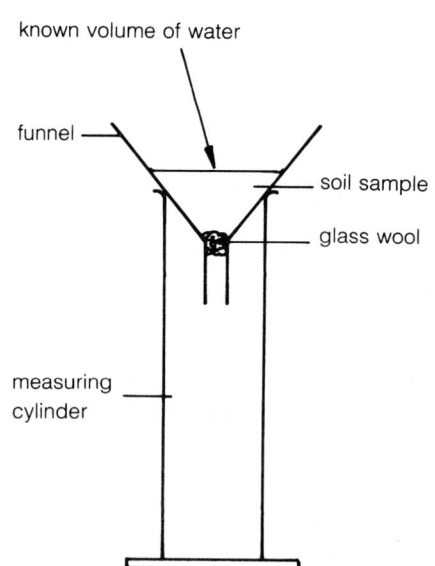

Use these tests to compare sandy, clay and loam soils, then construct a chart summarizing your results. These tests are especially useful when studying the relationship between soils and plant life in various habitats.

Investigating soil water

1 Water retention

Plug the neck of a filter funnel with a small wad of glass wool. Pour a known weight of dry, powdered soil into the filter funnel. Put the filter funnel on a measuring cylinder. Pour a known volume of water *gently* onto the soil and note the volume of water which eventually collects in the measuring cylinder. The difference between the two columns is the amount of water retained by the soil.

2 Rate at which water percolates through soils

After completing experiment **1** above, start a stop-watch and at the same moment pour another known volume of water onto the wet soil. Note the time taken for this water to drain through the soil.

3 Water content of soil

Take a sample of wet soil after completing experiment **2** above (note that it contains as much water as the soil can hold onto against gravity).

Weigh the sample, place it in an evaporating basin. Heat it in a thermostatic oven at 100 °C for 30 minutes and weigh it again.

Repeat until no further loss in mass has occurred. The difference between the original mass and the mass after heating is the mass of water in the soil. Convert this to a percentage of the original mass.

4 Questions:

Compare the *volume* of water retained by a soil sample (experiment **1**) with the *mass* of water it contains (experiment **3**). Are both results the same? (1 cm³ of pure water has a mass of 1 g.)

How do sand clay and loam soil differ in their ability to retain water and allow water to percolate through?

Why is water retention and percolation different in sand, clay and loam?

HAZARD WARNING

Take care when heating soil. WEAR EYE PROTECTION. Universal indicator is highly flammable and harmful if swallowed. KEEP AWAY FROM NAKED FLAME.

Your teacher will be looking for:	**and especially for:**
• careful use of the apparatus given	☐
• accurate measurement and presentation of results	☐
• accurate calculations	☐
•	☐

Comparing soil samples II

Measuring the humus, mineral and air content of soils

1 Humus and mineral content of soils
Weigh a sample of the dry soil after completing experiment **3**.
Place the sample in a crucible and heat it strongly for about 20 minutes.
Weigh it again when cool.

2 Questions:
Why was the soil burnt?
How can you calculate the mass of humus and minerals in soil from your results?
Calculate the humus and mineral content of your soil as percentages.
In what ways does adding humus to a soil increase its fertility?

3 Air content of soil
Find the volume of an empty tin. Punch small holes in the bottom of the tin and push it, open end first, into the soil until its bottom is level with the soil surface.
Dig the tin out of the soil *without* disturbing its contents and remove soil until it is level with the open end of the tin.
Empty *all* the soil into a large measuring cylinder containing a known volume of water. After bubbles have stopped rising, note the water level.

4 Questions:
How can you calculate the air content of soil from your results?
Do sand, loam and clay contain different amounts of air. If so, why?
Why is air an important part of soil?

Investigate soil particles and pH

Two very important features of soil are the amounts (**proportions**) of humus, silt, clay, sand and gravel that it contains and its pH. It is very easy to investigate these features with the following tests.

5 Soil particles
Half fill a measuring cylinder with soil, then fill to the top with water. Put your hand over the open end of the measuring cylinder and shake until soil and water are thoroughly mixed. Allow the soil to settle for at least five minutes.
Mineral particles settle in separate layers, the largest first. Humus floats to the top.
Use graduations on the measuring cylinder to estimate the proportions of each size of mineral particle in the soil, and the amount of humus.

6 Testing soil pH
Place a few drops of pH indicator on a white tile. Sprinkle a little soil into the indicator and mix thoroughly.
Tilt the tile so that the indicator runs out of the soil. Compare its colour with the chart provided with the indicator and read off its pH.

Your teacher will be looking for:	and especially for:	
• careful use of the apparatus given		☐
• accurate measurement and presentation of results		☐
• accurate calculations		☐
•		☐

HAZARD WARNING

WEAR EYE PROTECTION. BEWARE OF SPITTING HOT PARTICLES!

Further work
Find out:
- the names of some plants which grow best in acid soil and some which prefer alkaline soil;
- the types of habitats which are likely to have acid or alkaline soils (e.g. chalk grassland, moor, marsh, etc.);
- how you can change the pH of the soil.

Investigating the functions of soil microbes	BIOLOGY ACTIVITY	A41

You need:

large plant pots

safety goggles

graph paper

leaves from deciduous trees

soil

large tin

Bunsens and tripods

distilled water

1 Obtain two leaves such as oak, beech or sycamore (not thick leathery leaves, like holly). Attach a piece of paper to their petioles and label them 'Leaf A' and 'Leaf B'. Weigh each leaf. Place each leaf on graph paper and carefully draw around it. Label the drawings 'Leaf A' and 'Leaf B'. Count the graph paper squares covered by the leaves to work out their area.
Write the area and weight of each leaf on the drawing.

2 Obtain enough rich loam soil to fill two large plant pots. Put half the soil in a tin.
Sterilize the soil by heating it over a Bunsen flame until it is completely burnt.

3 Label two large plant pots A and B.
Pot A: Half fill with unburnt soil, lay leaf A flat on the surface and pour in more soil until the pot is full. Water the soil with *distilled* water.
Pot B: Half fill with burnt soil, lay leaf B flat on the surface and pour in more burnt soil until the pot is full. Water with *distilled* water.

4 After two weeks empty the pots, keeping the soil separate and intact, to retrieve the leaves. Carefully wash off any soil and blot dry with paper towels, without damaging the leaves.

5 Compare the leaves visually. Reweigh them and compare with the original masses. Draw their outlines on graph paper, marking any holes which have appeared. Label the drawings 'Leaf A (2 weeks)' and 'Leaf B (2 weeks)'.
Compare the area of each leaf with its original area.

6 Replant leaf A as before. Sterilize the soil from pot B again, then replant leaf B as before.

7 Repeat steps **4**, **5**, and **6** above at two-week intervals until there is a clear difference between the two leaves.

8 **Questions:**
What is the difference between the soils in pot A and pot B?
What changes occurred in the mass, and the area of each leaf?
What caused these changes?
How do these changes help maintain soil fertility?
'If it were not for the decay processes, demonstrated by this experiment, all plants and animals would die'. What is the reasoning behind this statement? Is it true?

Your teacher will be looking for: **and especially for:**
- accurate observation ☐
- good presentation of results ☐
- sensible conclusions ☐
- .... ☐

HAZARD WARNING

Take care when heating soil. WEAR EYE PROTECTION. BEWARE OF SPITTING HOT PARTICLES!

Ecology I – making transects

You need:

string	notebooks
poles (3 metres long)	graph paper
metre rules	plant identification books
spirit level	tent pegs

Making a line transect

Many habitats have areas where vegetation and animal life changes from one type to another, such as down a hillside or the banks of a stream. These changes can be studied by making a **line transect**. This is a record of the types of plant (and certain animals) which live on a line across a habitat.

1 Mark a length of string at 1 m or smaller intervals and stretch it between two pegs across the habitat so that it crosses an area of change.

2 On sloping ground another string can be stretched across the same line between two poles and arranged above the ground so that it is horizontal. At regular intervals along the slope measure the distance between the string and the ground. Use this information to draw an outline of the slope on graph paper.

3 Starting at one end of the transect, record the plants which touch the first string at certain intervals (e.g. 10, 50 or 100 cm, depending on the density of the vegetation).

4 Using symbols to represent each species, record the plants along the transect on graph paper (see example below).

Projects

5 Study the changes:
   - down the banks of a stream or pond
   - from the upper sea shore down to low tide level on rocky and other shorelines (show attached animals such as barnacles, limpets, mussels, etc.)
   - from a shaded to an unshaded area
   - across a path to study the effects of trampling on vegetation
   - up an overgrown wall from ground level
   - across a goalmouth on a grass pitch

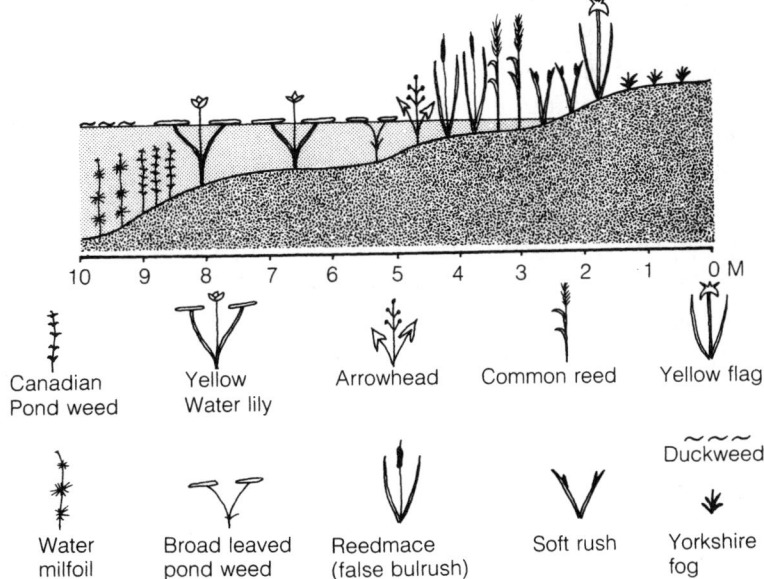

| 10 | 9 | 8 | 7 | 6 | 5 | 4 | 3 | 2 | 1 | 0 M |

Canadian Pond weed — Yellow Water lily — Arrowhead — Common reed — Yellow flag

Water milfoil — Broad leaved pond weed — Reedmace (false bulrush) — Soft rush — Duckweed — Yorkshire fog

6 Look carefully at the plants (and animals) along your transect and try to work out *why* there is a change from one type to another.
Does the soil pH level change?
Does the soil water content change?
Does the type of soil change (e.g. from clay to sand)?
Does one area get more light than another?
Is one area more exposed to wind than the other?

7 Look for adaptations to these and other changing conditions across a habitat.
How are the plants adapted to live in wet or dry conditions, or windy or sheltered conditions, or shaded or unshaded conditions, etc.?
Does animal life change as the vegetation changes? If so, in what ways?

Making a belt transect

This is a record of vegetation in a narrow strip, or belt, across a habitat. It is useful where a line transect fails to show enough plants to make a worthwhile record.

8 Peg out two parallel lines 1 metre apart and as long as you wish across a habitat. Study the area one square metre at a time by dividing it into 1 metre squares with string.

9 Choose symbols to represent one plant which grows in patches, or patches of several different plants which grow together (see example below).

10 Record your results by drawing the area to scale on graph paper.

Projects

11 A belt transect can be constructed instead of a line transect in each of the projects mentioned opposite.

12 Colonization
Belt transects can be used to study **colonization**, which is the appearance of plants and animals in a cleared area of a habitat.
Peg out two parallel lines 1 metre apart and 3 metres long on grassland. Dig out all vegetation from the central square metre, removing as many roots as you can but leaving as much soil as possible.
Record vegetation in the undisturbed part of the belt and, at monthly intervals, record plants growing in the cleared area. What do your results tell you about how one set of plants gives way to another in **ecological succession**?

13 Devise a method of studying the colonization of rock on a sea shore by sea weeds and animals.

B – Creeping buttercup
C – White clover
D – Daisy
Da – Dandelion
R – Ribwort plantain
T – Spear thistle
Mole hill
Bare ground

Timothy grass
Meadow Foxtail grass
Crested dogstail grass
Sheeps fescue grass
Cocksfoot grass

A belt transect across a grassland path showing the effects of trampling

Your teacher will be looking for:	and especially for:
• safe and careful work in the field	☐
• good observation and presentation of results	☐
• sensible conclusions which fit your results	☐
•	☐

You need:

25 cm quadrats

graph paper

enamel paint

paint solvent

fine paint brush

notebooks

identification books (plant and animal)

white pie dishes

specimen tubes

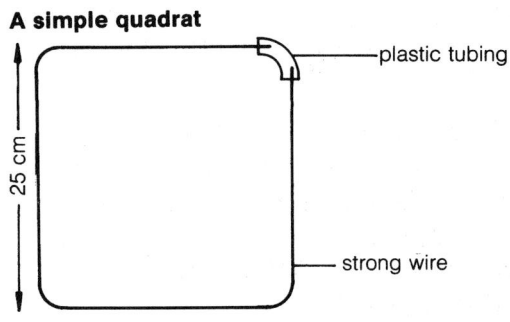

A simple quadrat

plastic tubing

25 cm

strong wire

Random sampling of plant life

How would you answer the question, 'Which types of plant are the most common in this habitat?' You could count all the different types of plant but this is not necessary except for very small habitats. An easier method is called **random sampling**. To do this you use a square or rectangular frame called a **quadrat** to study several small areas (**samples**) of the habitat chosen at random.

You place a quadrat at random throughout a habitat by throwing one over your shoulder. (CARE!) Do not deliberately throw it to land on vegetation which looks interesting. What you do next depends on the information you require. This could be the *density*, *frequency* or *percentage cover* of various types of plant.

1 **Density** This is the number of plants (or animals) in a unit area of habitat (e.g. the number per 25 centimetre square). To discover the density of a plant species in a habitat, you count the number of this species present inside the area of a quadrat each time it lands. Continue until the quadrat has been cast throughout the whole habitat, then calculate the average number of times the species was found.

2 **Frequency** This is the number of times that a particular species is found when a quadrat is thrown a certain number of times. To calculate frequency you count the number of different species within the quadrat each time it lands and note their names. If, for example, you throw it a hundred times, you note the number of times each species was found and express each result out of a hundred. This will tell you the most common (most frequent) species in the habitat, then the next most common, down to the rarest.

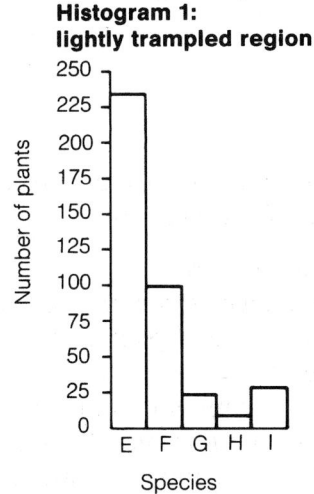

Histogram 1: lightly trampled region

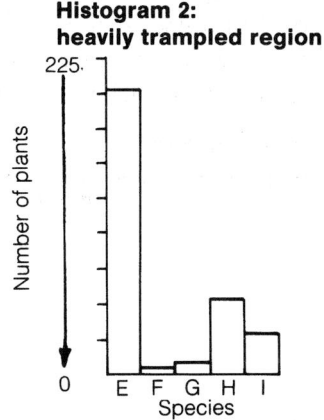

Histogram 2: heavily trampled region

(continued on page 84)

3 Percentage cover This is the percentage of the total area of a habitat which each species covers. This can only be calculated with species which grow in clumps or large patches. Estimate the percentage cover of the species found inside a quadrat each time it lands, then calculate the average cover over a number of throws.

Project

You can use these methods to compare different habitats.

4 Study the boundary between one habitat and the next to discover how plant frequencies change.
Make a list of plants from the most frequent to the rarest in each habitat.
Where possible find the percentage cover of the commonest plants in each habitat.
Present results as histograms.

Studying animal populations

One aim of studying animal populations is to list species from the most common to the rarest. This is difficult because animals move in and out of an area, but there are ways of solving this problem.

5 Comparative method Carefully search the habitat and compare the numbers of each species you find. Give each species a score: five for the most frequently met and one for the rarest. If several groups work independently, average scores can be obtained.

6 Capture-recapture method You can estimate the total population of species which can be caught easily and marked in some way, and which disperse quickly when released, e.g. dragonflies, large beetles, snails, crabs, woodlice, water boatman, etc.
Catch a number of specimens, mark them with brightly coloured enamel paint, then release them. After a day or two, when they have thoroughly dispersed, try to capture the same number again.
Note the number of marked specimens in the second catch.
An estimate of the total population of this species is calculated by this formula:

$$\text{Population} = \frac{\text{total in first catch} \times \text{total in second catch}}{\text{number of marked specimens in second catch}}$$

Question 1

The histograms on the opposite page show the number of plants found in random samples taken in lightly trampled and heavily trampled grassland.
Which species is most affected by trampling?
Which species is least affected?
Study trampled areas and try to find out which plants can live there.
Try to explain how some plants are adapted to withstand trampling.
Explain how you would obtain random samples like these.

Question 2

Some students caught 50 crabs in a 25 m² area of seashore, marked them with yellow paint, then released them. Four days later they caught another 50 crabs. Thirteen of these were marked with yellow paint. What is the estimated total crab population for this area? Why is this only a rough estimate of the crab population?

Your teacher will be looking for:	**and especially for:**	
• safe and careful work in the field		☐
• good observation and presentation of results		☐
• sensible conclusions which fit your results		☐
•		☐

HAZARD WARNING

Take care when 'throwing' quadrants.

1 Students **must** be supervised at all times.

2 Students (and teachers) **must** wear protective clothing and safety glasses.

3 There **must** be no hand to mouth activities: e.g. chewing pencils, licking labels, eating, smoking, drinking, mouth pipetting, etc.

4 Benches and floor areas **must** be kept free of bags, etc. *before* starting practical work.

5 All exposed cuts/abrasions **must** be covered with waterproof dressings/disposable gloves.

6 Benches should be swabbed immediately before and after each practical. Benches should be swabbed so they remain wet for a minimum of 10 minutes. Only provided disenfectant should be used. *Skin contamination only* should be swabbed with 70% alcohol.

7 Students and teachers should be fully conversant with asceptic techniques. These should be practised until they become habitual. Always work close to a Bunsen flame.

8 All micro-organisms should be treated as if they are pathogenic. *(They may have become contaminated or mutated.)*

9 Lids **must** not be placed on benches. They should be held by the little finger.

10 Swabs/specimens **must** not be taken from:
   • any orifice/human mucus, etc.
   • anywhere within toilet boundaries, drains, etc.
   • make-up brushes, etc.

   • isolated coliforms using McConkey agar/broth

Accidental occurences of these **must** be reported. Inform a technician.

11 Anaerobic cultures **must** be avoided. Take especial care when:
   • taping plates
   • streaking (agar that is broken during streaking constitutes anaerobic conditions)

12 The creation of 'aerosols' should be avoided:
   • Flamed loops may cause sputtering. This can be avoided by heating the base of the wire near the handle first and slowly withdrawing the rest of the loop just above the cone. Ensure all of the wire loop is red hot and the handle is held at an angle to prevent any contamination running down it.
   • Hot loops **must** not be plunged into broth, etc.

13 All spillages **must** be reported immediately.

14 Incubation should only take place at room temperature (max 30 °C).

15 Plates **must** not be opened once taped. (On occasions where this is essential, the micro-organisms **must** have been killed prior to the plates being returned to the student.)

16 Hands **must** be washed thoroughly with soap and water *before* leaving the laboratory – **on all occasions**.

17 Cultures and plates **must** be sterilized before disposal.

You need:

Petri dishes of sterile nutrient agar and tubes of nutrient broth

bacteria culture

incubator

various disinfectants

sterile bulb pipettes

Bacteria are all around us

1   Label five sterile Petri dishes of nutrient agar A to E.

2   **Dish A:** Take off the lid and expose the agar to air in the laboratory for an hour. Replace the lid and seal it with sticky tape.
    Dish B: Sprinkle a little dust from a bench surface or the floor onto the surface of the agar, replace the lid and seal it.
    Dish C: Add a few drops of rain water to the agar, replace the lid and seal it.
    Dish D: Add dead fly. Replace the lid and seal it.
    Dish E: Add a few drops of distilled water, replace the lid and seal it.

3   Incubate the Petri dishes upside down for at least 48 hours at 15 °C to 30 °C.

4   Count the number of bacteria and mould colonies on each plate.
    What is the function of dish E?
    What conclusions can you draw from these results?

Investigate the effects of disinfectant on bacteria

5   Prepare five test tubes of sterile nutrient broth and label them A to E.
    Add three drops of bacteria culture to each tube and replace the cotton wool plug.

6   **Tube A:** Add 5 cm³ of full strength disinfectant, replace the plug and seal it with sticky tape.
    Tube B: Add 5 cm³ of disinfectant diluted to $\frac{1}{10}$ of full strength, replace the plug and seal it.
    Tube C: Add 5 cm³ of disinfectant diluted to $\frac{1}{100}$ of its full strength, replace the plug and seal it.
    Tube D: Add 5 cm³ of disinfectant diluted to $\frac{1}{1000}$ of its full strength, replace the plug and seal it.
    Tube E: Add 5 cm³ of distilled water, replace the plug and seal it.

7   Incubate the tubes at 25 °C to 30 °C for at least 48 hours. Compare the cloudiness (**turbidity**) of each.
    What conclusions can you draw from the effect of disinfectant on bacterial growth?

Your teacher will be looking for: **and especially for:**

•   careful and safe use of the apparatus given   ☐
•   accurate observations   ☐
•   good presentation of results   ☐
•   sensible conclusions which fit your results   ☐
•   ....   ☐

HAZARD WARNING

Never culture material from a lavatory, sewage-polluted water, animal cage or any human source e.g. finger nail scrapings.

You need:

Petri dishes of sterile nutrient agar and tubes of nutrient broth

sterile bulb pipettes

tubes of nutrient broth

incubator

bacteria culture

warm water, soap, and paper towels

How clean are your hands?

1 Label four Petri dishes of sterile nutrient agar A, B, C, and D.

2 **Dish A:** Take off the lid for 10 seconds, then replace it and seal the rim with sticky tape.
Dish B: Take off the lid and press the fingers of an *unwashed* hand onto the agar (it must not be broken up by too much pressure). Replace the lid within 10 seconds and seal it.
Dish C: Wash your hands in warm water only (no soap). Dry them with paper towels. Take off the lid and touch the agar as before. Replace the lid within 10 seconds and seal it.
Dish D: Wash your hands *thoroughly* using warm water and soap. Dry them with paper towels. Take off the lid, touch the agar as before, replace the lid within 10 seconds and seal it.

3 Incubate the Petri dishes upside down at 37 °C for a week. *Without opening them:*

   a) count the number of bacteria colonies in each dish

   b) count the number of different colonies in each dish

   c) design a results table

   What do your results tell you about the cleanliness of washed/unwashed hands and the effectiveness of soap as a cleaning agent?

Investigate the effects of pH on bacterial growth

4 Add 8 cm^3 of sterile nutrient broth to each of five sterile boiling tubes marked A, B, C and D.

5 Add 1 cm^3 of 0.1M hydrochloric acid to tube A.
Add 1 cm^3 of 0.0001M hydrochloric acid to tube B.
Add 1 cm^3 of distilled water to tube C.
Add 1 cm^3 of 0.0001M sodium hydroxide to tube D.
Add 1 cm^3 of 0.1M sodium hydroxide to tube E.

6 Inoculate each tube with 1 cm^3 of bacterial culture, plug them with cotton wool and seal with sticky tape.
Incubate the tubes at 25 °C to 30 °C for 48 hours.
Compare the cloudiness (**turbidity**) of each tube. What do your results tell you about the effect of Ph on bacterial growth?

Further work

a) A hospital wants to test a new bactericidal soap for cleaning the hands of surgeons. Design an experiment to compare the new soap with the one they already use.

b) Onions and other vegetables can be preserved in weak acids such as vinegar. Design an experiment to find the strength of vinegar required to preserve vegetables at room temperature for at least one month.

Your teacher will be looking for: **and especially for:**

- careful and safe use of the apparatus given ☐
- accurate observation and recording of results ☐
- good presentation of results in tables ☐
- sensible conclusions ☐
- .... ☐

HAZARD WARNING

Wash hands thoroughly after touching agar. During incubation, plates SHOULD NOT be completely sealed. After incubation, plates must be completely sealed with tape before observing results.

Investigating soil life

Soil contains living things

1 Predict what will happen to liquid in the U-tube of the apparatus below when the taps are closed. (**Clue:** Living things take in oxygen and produce carbon dioxide at about the same rate. Carbon dioxide is absorbed by the soda lime.)

2 Prepare the apparatus, leave the taps open for at least five minutes, then close them and check your prediction.

3 **Questions:**
Why does the coloured liquid move in the way you observed?
What exactly is the apparatus measuring?
Why was the sterilized soil included?

Demonstrate the presence of microbes in soil

4 Sprinkle a *little* soil from a number of different sources into dishes of sterile nutrient agar. Label them and seal with sticky tape.

5 Incubate the dishes, upside down at about 25 °C. Examine the dishes daily for growths of bacteria and fungi.

6 Compare the types of mould and bacterial colonies which appear from each soil sample. (*Do not open the dishes. Observe the safety precautions.*)

7 Design two controls for this experiment to prove that the microbes come only from the soil and not from the air or from contaminated Petri dishes.

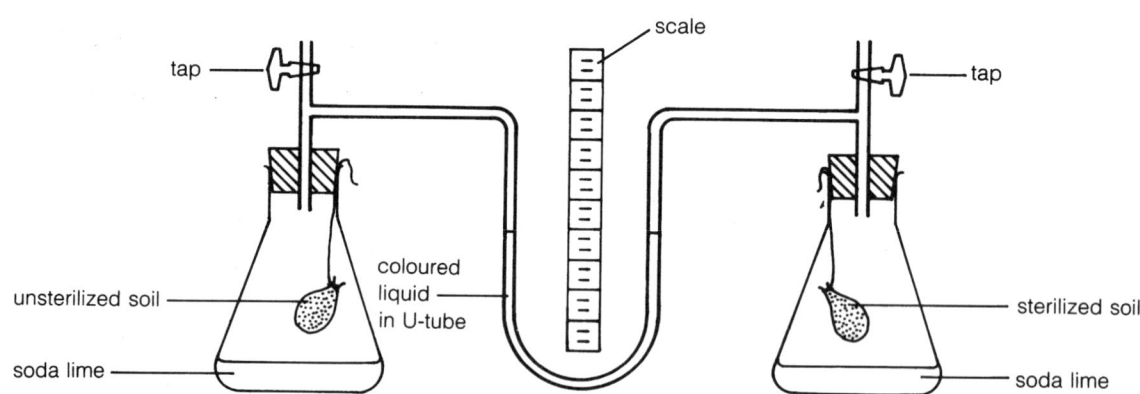

Your teacher will be looking for: and especially for:

- careful and safe use of the apparatus given ☐
- accurate observation ☐
- good presentation of results ☐
- sensible conclusions ☐
- good design of controls ☐
- .... ☐

HAZARD WARNING

During incubation, plates SHOULD NOT be completely sealed. After incubation, plates must be completely sealed with tape before observing results.

Further work

Design an experiment to find out if heavy doses of inorganic fertilizer and various pesticides affect soil life. What factors must stay the same? What controls will be needed? How will you treat the soil samples?

You need:

sterilized test tubes plugged
  with sterile cotton wool

Bunsen, tripod and gauze

incubator

nutrient broth

bacterial culture

refrigerator

wire loops

500 cm³ beakers

Investigate the effects of cold and warm temperatures on bacteria

1 Add a few drops of bacteria culture to a bottle of nutrient broth.
Pour 2 cm³ of the broth into each of three sterilized test tubes, plug each with sterile cotton wool and label the tubes A, B, and C.

2 Incubate the tubes as follows:
Tube A: at 4 °C (the average temperature of a domestic refrigerator)
Tube B: at room temperature
Tube C: at about 60 °C

3 Observe the tubes daily for signs of cloudiness (**bacterial growth**).

4 What conclusions can you draw from your results about the effects of temperature on the growth of bacteria?

Investigate the effects of high temperatures on bacteria

5 Pour 2 cm³ of nutrient broth into each of six sterile test tubes, add a few drops of bacterial culture, then plug them with sterile cotton wool.
Leave the tubes in a warm place for a day.

6 Call the next day 'day one'.
Day one: Boil all six tubes in a beaker of water for five minutes, then put them in a warm place.

Day two: Put two tubes on one side in a warm place and label them A and B. Boil the other four tubes for five minutes and put them in a warm place.
Day three: Put two tubes on one side in a warm place and label them C and D. Boil the other two for five minutes, then label them E and F. Put them with the others in a warm place.

7 **Questions:**
How many times have each pair of tubes been boiled?
What effect does this have on the growth of bacteria inside them?
Boiling kills bacteria but not their reproductive spores. How does this explain your results?

8 **Find out** if results are different when this method is repeated but this time the tubes are placed in a pressure cooker for five minutes instead of boiling.
What is the difference between pressure cooking and boiling?
Does this affect bacterial spores?

Your teacher will be looking for: **and especially for:**

- careful and safe use of the apparatus given ☐
- accurate observations ☐
- good presentation of results ☐
- sensible conclusions ☐
- .... ☐

HAZARD WARNING

Do not remove cotton wool plugs at any time during this activity.

PART C

Teacher's Notes

This section contains 12 Investigation sheets and 6 controlled Assessment Tasks. Other topics suitable for open-ended investigations are listed. In addition, there is a set of photocopiable notes for students about the steps involved in carrying out investigations. All these materials are designed to assist teachers in assessing the experimental and investigative skills required by the national curriculum.

Science 1: Each of the Activities detailed here provides ample opportunity for the practice and assessment of Sc1. Each Activity has been coded P, O, A, or E to assist teachers in identifying appropriate Sc1 opportunities To help further with this task each of the check boxes contains an open section which the teacher can use to define precisely the specific aspect of Sc1 he or she wishes a student to focus on.

P – planning experimental procedures

O – obtaining evidence

A – analysing evidence and drawing conclusions

E – evaluating evidence

Investigation	HAZARD WARNING	Science 1			
		P	O	A	E
I1		✔	✔	✔	✔
I2		✔	✔	✔	✔
I3		✔	✔	✔	✔
I4		✔	✔	✔	✔
I5		✔	✔	✔	✔
I6		✔	✔	✔	✔
I7		✔	✔	✔	✔
I8		✔	✔	✔	✔
I9		✔	✔	✔	✔
I10		✔	✔	✔	✔
I11		✔	✔	✔	✔
I12		✔	✔	✔	✔

Assessment Task	HAZARD WARNING	Science 1			
		P	O	A	E
T1		✔	✔		
T2		✔	✔	✔	
T3		✔	✔	✔	✔
T4		✔	✔	✔	
T5		✔	✔	✔	✔
T6			✔	✔	✔

Student's Notes: Designing an investigation ...

What is an investigation?

When you are given an experiment to do, your teacher may give you detailed instructions. You may be told what apparatus to use, how to set it up, and what to measure. However, when you are given a problem to **investigate** you have to decide what things to look at, how to measure them, and how to present your results. You also have to look closely at your results to decide whether or not you have solved the problem. **Investigating is an important scientific skill.**

Planning your investigation

- **Read the problem to make sure that you understand it.**

  Ask yourself, *'What am I trying to find out?'* Think, *'Have we studied any topics which might be connected to this problem?'* For example, growth of a plant may have something to do with photosynthesis and/or the effect of fertilisers, the tiredness we feel in our muscles may have something to do with anaerobic respiration, etc.

  At this stage you may be able to make a **prediction** or **hypothesis** which you can test through your experiment. Always try to support your prediction with your knowledge of science. It will help if you write, *'I predict that .... will happen because ....'*

- **Decide which things (factors) are involved in the problem.**

  Ask yourself, *'What factors am I going to investigate?'* For example, temperature, mass, time, size, colour, etc.? Then decide which of these you are going to change during the experiment. These are called **variables**. Make a note of the variables you are interested in.

- **Decide which factors shouldn't change during the investigation.**

  Ask yourself *'What things must I keep constant during my experiments?'* Remember, you can't handle too many changes at once. Concentrate on changing one variable at a time, keeping the others constant. If you have several factors to investigate, you may have to do more than one experiment. For example, if you are investigating the germination of seeds, you may have to do three separate experiments: **i)** changing the light intensity, **ii)** changing the temperature, **iii)** changing the amount of water.

- **Design any 'controls' that you need.**

  Many biology experiments need controls. These are parallel experiments which allow you to check that your results really are due to the factor being investigated. For example, imagine that you are asked to find out if spraying a crop with pesticide will kill water weed in a nearby pond. You put water weed in a test tube, add pesticide and it dies. But would it have died in a test tube without pesticide? You need a **control** – another

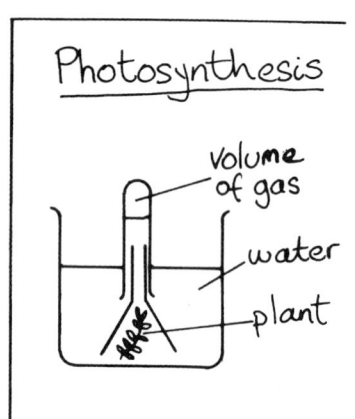

Photosynthesis

volume of gas

water

plant

factors

plant – fixed
(Elodea)

size – fixed
(measure length/mass + count leaves)

light – fixed
(on same window sill)

water temp. – change
(0°C – 40°C?)

vol. of gas – measure

92

plant of identical size and health, kept at the same amount of light, and nutrients, etc., but without the pesticide. In a controlled experiment all, the conditions are kept the same *except* for the one factor being investigated.

- **Make a plan of what you are going to do.**

  Write a description of what you're going to do. Your plan needn't be very long but be sure that it covers the following:
  the problem you are going to investigate
  a list of the apparatus you are going to need
  a labelled diagram of your experiment
  the measurements you are going to make
  how you are going to make sure that the test is fair
  what your predictions are

Carrying out your investigation

- **Follow your plan step by step.**

  Make a note of any problems you come across. You may have to change your plans to improve your investigation – make a note of these.

- **Put all your readings in a table.**

  Tables are the best way of recording your measurements. Examples of tables are shown on the right. Remember, when taking readings:
  do use headings in your table
  do include the units
  do take at least six pairs of readings if you are going to plot a graph
  do write your readings straight into your table (don't write them on scrap paper)
  do repeat your measurements if possible

Presenting evidence and drawing conclusions

- **Include all your readings.**

  All your readings are important – even those that do not seem to 'fit'. If you think that one of your readings is not 'right', make a note of it. For example, if a reading does not fall close to all the others you may write, *'The third reading for the temperature seems too large. This may be because I didn't stir the water before taking the temperature. I repeated this measurement twice and the other readings were close together so I used these to find an average.'*

- **Show any calculations you have done.**

- **Include any graphs you have plotted.**

  Many, but not all, investigations need graphs to show patterns in results. When you take pairs of readings, you *choose* which values to give one of the quantities – like choosing to measure the time at 1, 2, 3 minutes and so on. Usually, this quantity (the independent variable) goes along the bottom axis. The quantity which you measure (the dependent variable) goes up the side axis. Remember to label both axes of each graph.

You may need a table for your readings like this:

mass of peanut = ... g

volume of water = ... cm³

starting temp. = ... °C

final temp. = ... °C

If you have several sets of readings you will need a table like this:

length of seedling in mm	time in days
4	1
4.3	2
4.9	3

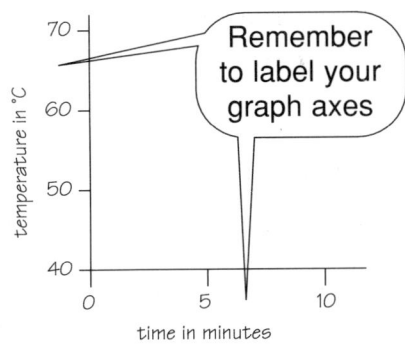

- **Use your results to draw conclusions.**

  Look at your results and decide what they show. For example, your graphs may show a pattern between the variables you have been investigating. A *straight line* graph shows a fairly simple relationship. However, *curved* graph lines are equally important and you should be able to explain what you have found. Remember, your conclusions must fit what you have found.

- **Explain your conclusions.**

  When you have come to one or more conclusions, try to support them using your knowledge of science. Try to use laws, theories, or models you have studied to explain your results.

- **Evaluating your investigation**

  The final part of any investigation is to evaluate your work. This means thinking about the things that went well and the things which could be improved. Some of the things you might include are:

1 How good were your results? For example, you may be able to compare your values with those given in textbooks or data books.

2 How likely is it that you would get the same results if you did it again? For example, if you were investigating genetic inheritance, were you able to take enough readings to feel confident that your results were 'typical'?

3 Does your conclusion confirm your original prediction? Or, have the results disproved your prediction/hypothesis?

4 How would you improve your investigation if you were to do it again? Are there any other factors which you should have taken into account? Are there any *further* experiments which you need to do to support your conclusions?

> My results do fit my hypothesis but I only used one type of water weed. I would like to repeat my experiment with a range of plants found in British ponds.

Light and photosynthesis

When plants photosynthesise, they produce oxygen. When water plants give off oxygen, small bubbles of the gas can be seen. A simple way of measuring the rate of photosynthesis is to count the bubbles released in a certain period of time.

Design and carry out a scientific experiment to find out whether the intensity of light affects the rate of photosynthesis in the pond weed *Elodea*.

'ello dear

- ♦ **Start by thinking about the following:**

- What do you know about photosynthesis?

- What apparatus will you need to study photosynthesis in *Elodea*?

- How will you vary light intensity? Will you have to set up any controls?

- What do you *predict* will happen? Why?

- Are there any *Safety Hazards*?

- What safety precautions **must** you take?

- ♦ **Plan your investigation**

- ♦ **Let your teacher check your plans**

- ♦ **Carry out your investigation**

- ♦ **Write up:**

- what you did(include diagrams)

- what you found (your results including any tables and/or graphs)

- what your conclusions are and how they relate to your predictions

- how your experiment could be improved

Your teacher will be looking for:	**and especially for:**
• the use of a sensible scientific method	☐
• careful observations and measurements	☐
• good presentation of results	☐
• a sensible conclusion	☐
• sensible suggestions about improving the experiment	☐
•	☐

Leaf colour and photosynthesis	BIOLOGY	I2
	INVESTIGATION	

Most plants are green and you have probably done experiments to show photosynthesis in green leaves. However, some plants have red or purple leaves, others have silvery-grey foliage. There are also plants where some parts of the leaf are white and other parts are green (variegated).

Design and carry out an investigation into photosynthesis in different coloured leaves.

♦ **Start by thinking about the following:**

• What do you know about photosynthesis?

• What apparatus will you need and how will you use it?

• What type of leaves will you use?

• How will you prepare the leaves for testing?

• Write down any *hypotheses* you are going to test.

• What safety precautions **must** you take?

♦ **Plan your investigation**

♦ **Let your teacher check your plans**

♦ **Carry out your investigation**

♦ **Write up:**

• what you did (including diagrams)

• what you found (your results including any tables and drawings)

• what your conclusions are

• whether your hypotheses were supported (proved) or not

• any scientific explanation you can offer for your conclusions

• how your investigation could be improved

Your teacher will be looking for:	**and especially for:**
• **safe** use of a sensible method	☐
• use of a good variety of leaves carefully prepared	☐
• careful observations	☐
• good presentation of results	☐
• sensible conclusions	☐
• sensible suggestions about improving the experiment	☐
•	☐

| Transpiration through leaves | BIOLOGY INVESTIGATION | I3 |

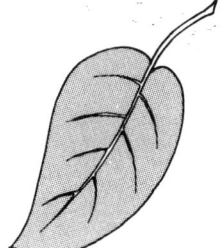

When plants transpire, water is lost through their leaves.

Design and carry out an investigation to decide whether transpiration takes place through the upper surface of the leaf, the lower surface, or both!

♦ **Start by thinking about the following:**

- What do you know about transpiration?

- How will you detect water loss?

- What apparatus will you need and how will you use it?

- How many leaves will you test? Will they need to be on the plant or not?

- Write down any *hypotheses* you are going to test.

- Are there any *Safety Hazards*?

- What safety precautions **must** you take?

♦ **Plan your investigation**

♦ **Let your teacher check your plans**

♦ **Carry out your investigation**

♦ **Write up:**

- what you did (including diagrams)

- what you found (your results including any tables)

- what your conclusions are

- whether your hypotheses were supported (proved) or not

- any scientific explanation you can offer for your conclusions

- how your investigation could be improved or extended

Your teacher will be looking for:	**and especially for:**
• use of a sensible method	☐
• use of a sensible number of leaves carefully prepared	☐
• careful observations and measurements	☐
• good presentation of results	☐
• sensible conclusions which match your results	☐
• sensible suggestions about improving the experiment	☐
•	☐

Investigating muscle fatigue and recovery	BIOLOGY *INVESTIGATION*	I4

If you hold a heavy object at arm's length you soon get tired. It becomes impossible to keep your arm straight. This is because the muscles get tired. They suffer *fatigue*.

If you rest for a short period, your muscles will recover and you can hold the weight out again – but this time you may not be able to keep your arm straight for so long!

Investigate the effect of different rest periods on muscle recovery.

♦ **Start by thinking about the following:**

• What do you know about muscle fatigue? What is happening in the muscles?

• How are you going to measure the 'strength' or condition of the muscles before and after rest?

• What do you *predict* will happen? What scientific knowledge have you used to make this prediction?

• Will you work on your own or will you use other people?

• What range of rest periods will you use?

• How will you display your results?

• Are there any *Safety Hazards*?

• What safety precautions **must** you take?

♦ **Plan your investigation**

♦ **Let your teacher check your plans**

♦ **Carry out your investigation**

♦ **Write up:**

• what you did

• what you found (your results including any tables and graphs)

• what your conclusions are

• whether your prediction was right or not

• how your investigation could be improved

Your teacher will be looking for:	and especially for:
• a sensible experimental design	☐
• a fair test of muscle condition	☐
• careful measurements	☐
• good presentation of results	☐
• sensible conclusions	☐
• sensible suggestions about improving the experiment	☐
•	☐

| Investigating respiration rates | BIOLOGY INVESTIGATION | I5 |

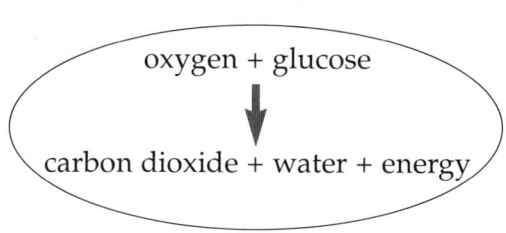

When animals respire aerobically, they use oxygen (usually from the air) and produce carbon dioxide gas. By monitoring how fast they use oxygen (and produce CO_2) we can measure their respiration rate.

Design and carry out an investigation into the effect of temperature on the respiration rate of maggots or other invertebrates.

oxygen + glucose
↓
carbon dioxide + water + energy

♦ **Start by thinking about the following:**

• Have you seen any apparatus suitable for measuring respiration rates? If not, try to find a suitable method in a textbook (or this book!).

• What range of temperatures will you use? (Remember you must not injure or kill the maggots.) How will you control the temperature?

• Are there any other factors which you must keep constant?

• What do you *predict* will happen? Write down any *hypotheses* you are going to test.

• Are there any *Safety Hazards*?

• What safety precautions **must** you take?

♦ **Plan your investigation**

♦ **Let your teacher check your plans**

♦ **Carry out your investigation**

♦ **Write up:**

• what you did (including diagrams)

• what you found (your results including any tables)

• what your conclusions are

• whether your hypothesis was supported (proved) or not

• any scientific explanation you can offer for your conclusions

• how your investigation could be improved

Your teacher will be looking for:	and especially for:
• use of a sensible method	☐
• use of a sensible range of temperatures carefully controlled	☐
• careful observations and measurements	☐
• good presentation of results	☐
• sensible conclusions	☐
• sensible suggestions about improving the experiment	☐
•	☐

Fitness and lung volume	**BIOLOGY** *INVESTIGATION*	**I6**

Athletes have to produce energy quickly. It helps if they can get oxygen to their muscles as it is needed. If not, an oxygen debt builds up, the muscles start to respire anaerobically and fatigue sets in.

Design and carry out an experiment to see if there is a relationship between lung volume and 'fitness'.

♦ **Start by thinking about the following:**

• How can you measure lung volume?

• What do you mean by fitness? Is it about power or is it about recovery rate? Or can you think of a better definition?

• How many people will you need to test?

• Are there any other factors you should take into account?

• Write down any *hypotheses* you are going to test.

• Are there any *Safety Hazards* or hazards to health?

• What safety precautions **must** you take?

♦ **Plan your investigation**

♦ **Let your teacher check your plans**

♦ **Carry out your investigation**

♦ **Write up:**

• what you did (including diagrams)

• what you found (your results including any tables)

• what your conclusions are

• whether your hypotheses were supported (proved) or not

• any scientific explanation you can offer for your conclusion

• how your investigation could be improved

Your teacher will be looking for:	**and especially for:**
• use of a sensible method for measuring fitness	☐
• use of a sensible method for measuring lung volume	☐
• careful observations and measurements	☐
• good presentation of results	☐
• sensible conclusions	☐
• sensible suggestions about improving the experiment	☐
•	☐

Investigating artificial meat

Many people choose not to eat meat. Some do not think that it is healthy and others do not like the way animals bred for meat are treated.

However, it is now possible to buy 'artificial meat'. Some is made from the soya bean plant *(Glycine Soja)*. Another type is made from a fungus called *Fusarium* which can grow on potatoes, starch, or wheat. The artificial meat is processed to make it look, feel, and taste like real meat.

Design and carry out an investigation to compare real meat with artificial meat.

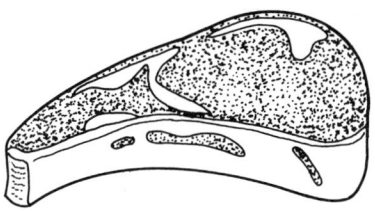

Real or artificial?

♦ **Start by thinking about the following:**

• What can you find out about artificial meat?

• What do you know about food tests? What food groups will you test for in this investigation?

• How are you going to make sure that your tests are 'fair'?

• How will you display your results?

• Are there any *Safety Hazards*?

• What safety precautions **must** you take?

♦ **Plan your investigation**

♦ **Let your teacher check your plans**

♦ **Carry out your investigation**

♦ **Write up:**

• what you did

• what you found (your results including any tables)

• what your conclusions are

• how your investigation could be extended to include factors such as taste

Your teacher will be looking for:	and especially for:
• a sensible experimental design	☐
• safe use of apparatus and chemicals	☐
• careful and accurate observations	☐
• good presentation of results	☐
• sensible conclusions	☐
• sensible suggestions about extending the experiment	☐
•	☐

Improving garden soil

Some plants grow better in soils which hold the water so that the roots are always surrounded by moist earth. Others prefer quick draining soils.

Gardeners try to make ideal soils for a wide range of plants by mixing ordinary garden soil with sharp sand and/or peat.

Design and carry out an investigation into the effect of mixing garden soil with:

a) sharp sand or grit

b) peat or coir

Cacti like free-draining soils

Ferns like moist soils

♦ **Start by thinking about the following:**

• What do you know about testing soil for 'water retention' and 'permeability'?

• What apparatus will you need and how will you use it?

• How many mixtures of soil will you test? How will you control their composition?

• Write down any *hypotheses* you are going to test.

• Are there any *Safety Hazards*?

• What safety precautions **must** you take?

♦ **Plan your investigation**

♦ **Let your teacher check your plans**

♦ **Carry out your investigation**

♦ **Write up:**

• what you did (including diagrams)

• what you found (your results including any tables)

• what your conclusions are

• whether your hypotheses were supported (proved) or not

• any scientific explanation you can offer for your conclusions

• how your investigation could be improved

Your teacher will be looking for:	and especially for:
• use of a sensible method	☐
• use of a sensible range of soils careful prepared	☐
• careful observations and measurements	☐
• good presentation of results	☐
• sensible conclusions	☐
• sensible suggestions about improving the experiment	☐
•	☐

Investigating plants and soil acidity	BIOLOGY INVESTIGATION	I9

Some plants grow better in soils which are slightly acid (pH<7) whilst others prefer slightly alkaline soils.

First do some research to find the names of common garden plants which are 'acid-haters' and those which are 'acid-lovers'.

Then find examples of those plants growing in gardens or even pots. Investigate the acidity of the soil in which they are growing in order to test the information you obtained in your research.

Garden Hint

Cabbages and other brassicas prefer a slightly alkaline soil. Prepare the vegetable bed by adding a light dusting of lime before planting.

♦ **Start by thinking about the following:**

• What do you know about testing soil pH? What will you need?

• Write down the *hypothesis* you are going to test for each plant.

• Who may be able to help you find suitable plants?

• How are you going to decide whether a plant is growing well or not?

• How many plants will you need to investigate to test your hypotheses?

• Are there any *Safety Hazards*?

• What safety precautions **must** you take?

♦ **Plan your research and investigation**

♦ **Let your teacher check your plans**

♦ **Carry out your research and investigation**

♦ **Write up:**

• what you did

• what you found (your results including any tables)

• what your conclusions are

• whether all your hypotheses were supported (proved) or not

• how your investigation could be improved and/or tested in the laboratory

---

Your teacher will be looking for: **and especially for:**

• a sensible choice of plants ☐
• careful observations and measurements ☐
• good presentation of results ☐
• sensible conclusions ☐
• sensible suggestions about improving the experiment ☐
• .... ☐

Investigating fermentation

Fermentation uses yeast to convert sugars into alcohol. It is one of the oldest uses of biotechnology. As early as 6000 BC the Babylonians were brewing beer and by 4000 BC the Egyptians were using yeast to make bread rise.

Design and carry out an investigation into the factors affecting the rate at which fermentation takes place.

Fermentation – useful biotechnology

♦ **Start by thinking about the following:**

• What do you know about fermentation?

• What apparatus will you need and how will you monitor the rate of fermentation?

• What factors will you investigate? How will you keep other factors constant?

• How many experiments will you need to carry out?

• Write down any *hypotheses* you are going to test.

• Are there any *Safety Hazards*?

• What safety precautions **must** you take?

♦ **Plan your investigation**

♦ **Let your teacher check your plans**

♦ **Carry out your investigation**

♦ **Write up:**

• what you did (including diagrams)

• what you found (your results including any tables and graphs)

• what your conclusions are

• whether your hypotheses were supported (proved) or not

• any scientific explanation you can offer for your conclusions

• how your investigation could be improved

Your teacher will be looking for:	and especially for:
• use of sensible methods	☐
• good choice of factors to investigate	☐
• careful observations and measurements	☐
• good presentation of results	☐
• sensible conclusions which fit your results	☐
• sensible suggestions about improving the experiment	☐
•	☐

How safe to refrigerate?

Many foods have to be kept in a refrigerator to stop them going off too quickly. Even in a refrigerator, food can only be kept for a certain length of time before it is unsafe to eat. Sometimes this information is given on the food label.

Design and carry out a scientific experiment to find out how long it is safe to keep food in a domestic fridge.

You should investigate food kept in the fridge all the time and food taken out for half an hour each day and then put back again (as could happen, for example, with cooked meat).

> KEEP REFRIGERATED
> AFTER OPENING
> EAT WITHIN FOUR DAYS

♦ **Start by thinking about the following:**

- Why does food go bad? Why does this make it unsafe to eat?

- What have you studied about 'bacteria'?
  Can you use this knowledge?

- What types of foods do you think will be best to use?

- What do you *predict* will happen? Why?

- How will you check for bacteria?
  (Visible checking is not good enough.)

- What safety precautions must you take?

- Are there any *Safety Hazards*?

- What safety precautions **must** you take?

♦ **Plan your investigation**

♦ **Let your teacher check your plans**

♦ **Carry out your investigation**

♦ **Write up:**

- what you did (include diagrams)

- what you found (your results including any tables)

- what your conclusions are and how they relate to your predictions

- how your experiment could be improved

Your teacher will be looking for:	and especially for:
• the design of a fair test	☐
• the use of proper safety precautions	☐
• careful observations and measurements	☐
• good presentation of results	☐
• a sensible conclusion	☐
• sensible suggestions about improving the experiment	☐
•	☐

Browning of apples and pH	**BIOLOGY** *INVESTIGATION* **I12**

Freshly cut apples gradually turn brown when they are left in air. Cooks try to stop this by dipping the slices of apple in lemon juice.

Design and carry out an investigation to find out whether pH affects the rate of the 'browning' reaction.

♦ **Start by thinking about the following:**

• How will you prepare solutions with a range of pH values?

• What apparatus will you need and how will you use it?

• How will you decide when the apple pieces have turned brown?

• Will you need to set up any controls?

• Write down any *hypotheses* you are going to test.

• Are there any *Safety Hazards*?

• What safety precautions **must** you take?

♦ **Plan your investigation**

♦ **Let your teacher check your plans**

♦ **Carry out your investigation**

♦ **Write up:**

• what you did (including diagrams)

• what you found (your results including any tables)

• what your conclusions are

• whether your hypothesis was supported (proved) or not

• any scientific explanation you can offer for your conclusions

• how your investigation could be improved

Your teacher will be looking for: **and especially for:**

• use of a sensible method ☐

• use of solutions with a sensible range of pH values ☐

• careful observations and measurements ☐

• good presentation of results ☐

• sensible conclusions ☐

• sensible suggestions about improving the experiment ☐

• .... ☐

Suggestions for further investigations

Arm strength
Test the following hypothesis: the strength of the human arm is directly related to body mass.

Testing eyesight
Design and carry out an investigation to see if sight is the same in both eyes.

Eyesight and age
Design and carry out an investigation into the effect of ageing on eyesight.

Testing hearing
Design and carry out an investigation to see if hearing is the same in both ears.

Hearing and age
Test the statement 'as you get older your hearing range changes'.

Toothpastes
Design and carry out an investigation into the contents, structure, and pH of a range of tooth cleaning products.

Teeth and cola
Design an investigation into the effects of 'cola' drinks on teeth.

'Junk food'
Does junk food really deserve its name? Investigate.

Exercise and body temperature
Investigate the effect of vigorous exercise on external and internal body temperature.

Products of respiration
Design an investigation to show that both aerobic and anaerobic respiration release energy in the form of heat.

Cleaning kitchen surfaces
Kitchen work surfaces may be cleaned with water, detergent, or disinfectant. Design an investigation into the effectiveness of these methods. *Safety: when growing microbe cultures, follow the safety rules.*

Mowing the grass
'Mowing a patch of lawn regularly changes the population of plants growing there'. Design and carry out an investigation to test this statement.

Selective herbicides
Some herbicides (weed killers) for use on lawns are said to kill 'broad-leafed' weeds only and to leave grass unharmed. Design an investigation into the effectiveness of such a weed killer. *Safety: weed killers are poisonous – take precautions.*

Herbicides and pond life
Design an experiment to test the statement 'herbicides washed out from farmers' fields can damage pond plants, even in low concentrations'. *Safety: weed killers are poisonous – take precautions.*

Germinating barley seeds
Design and carry out an experiment to show that barley seeds contain starch and that this is converted to sugar during germination.

Temperature and germination
Study the effect of temperature on the germination of seeds.

Factors affecting germination
'Seeds germinate more quickly if they have been soaked in water before planting'. Design an experiment to test this statement.

Flowers and insects
Design and carry out an investigation to see if insects are attracted more by the scent than by the colour of flowers.

Caterpillar feeding
Design and carry out an investigation into how much food a caterpillar eats in a day.

Winged seeds
Design and carry out an investigation into the effectiveness of 'wings' as a means of dispersing seeds from trees.

Colonisation
Investigate the colonisation of a 'bare' strip of soil.

Seed production in plants
Investigate the number of seeds produced by a single plant for a range of species. For example, how many seeds does a dandelion plant produce?

Fertilisers and plant growth

Design and carry out an experiment to compare the effectiveness of two commercial plant foods. *Safety: some fertilisers are poisonous – take precautions.*

Hormone rooting powder

Does hormone rooting powder really make plant cuttings root more quickly? Design an investigation to find out. *Safety: hormone rooting powder is harmful – follow the instructions – take precautions.*

Observing plant cells in solutions

Teacher/technician notes

Skills
This experiment can be used to assess a student's ability in:

- using apparatus (microscope)
- observing accurately
- presenting results as diagrams
- drawing conclusions

Previous lessons
Students should be familiar with the use of microscopes including making observations under high power.

Students should know about the structure of plant cells and about osmosis.

Apparatus and equipment
Each student should have:

- access to a microscope
- microscope slides and cover slips
- stick of rhubarb (or piece of onion)
- test tubes and rack
- dropping pipette
- strong sugar solution

This experiment works better if cells from the outer skin of rhubarb are used since these are easier to see without staining. However, some teachers may wish to use onion.

Procedure
Students will prepare slides of rhubarb cells in water and in sugar solution. They will observe and draw these under high power magnification.

Students will be asked to use their scientific knowledge to explain their results.

Cue sheets
No cue sheets are provided for this assessment task.

Criteria

Using apparatus and equipment

a) correct use of microscope

b) good preparation of slides

Making observations and measurements

a) cells in water are well observed

b) cells in sugar solution are well observed

Presenting data

a) clear, accurate diagram of cells in water

b) clear, accurate diagram of cells in sugar solution

Drawing and explaining conclusions

a) relates observations to movement of solute (water)

b) explains observations in terms of osmosis

Background information
There is no background information sheet for this Assessment.

Observing plant cells in solutions

Students' instructions

> **You need:**
>
> microscope   slides and cover slips   mounting needle
>
> dropping pipette   test tube and rack   scalpel
>
> strong sugar solution   stick of rhubarb

1 In this experiment you will observe the behaviour of plant cells in different liquids.

2 Prepare a thin strip of cells from the outer skin of the rhubarb. If you snap the stick of rhubarb, you may see the thin, outer 'skin' of cells. This can be peeled off.

3 Put drops of distilled water on one slide and strong sugar solution on another. (Label them so you know which is which!)

4 Put a piece of rhubarb skin on each slide and then gently lower coverslips over them. If your slides have too many air bubbles in them, start again.

5 Observe the cells under low, medium, and then high power.

6 Describe and draw any changes which you observe in the cells. Do this for the cells in water and those in sugar solution.

7 Explain your observations using your scientific knowledge.

8 Remove the cover slip from the cells in sugar solution. Wash the strip of cells and then make a new slide by mounting them in water.

9 Observe what happens and record your results.

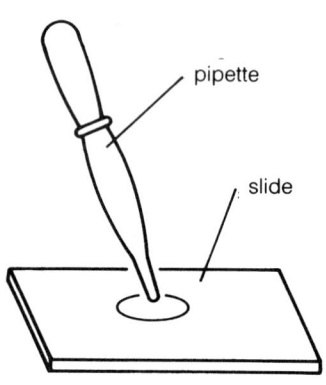

pipette

slide

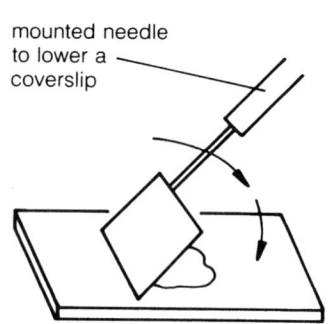

mounted needle to lower a coverslip

This experiment will test your skills in:	and especially in:
• using apparatus and equipment	☐
• observing carefully and accurately	☐
• presenting results in a table	☐
• drawing and explaining conclusions	☐
•	☐

HAZARD WARNING

Scalpels are sharp, handle with care.

Investigating photosynthesis in leaves

Teacher/technician notes

Skills
This experiment can be used to assess a student's ability in:

- using apparatus and equipment safely
- observing accurately
- presenting results in a table
- drawing conclusions

Previous lessons
Students should have studied photosynthesis.

They should also be familiar with the test for starch.

Apparatus and equipment
Each group should have:

- access to a plant kept in the dark
- access to a plant kept in the light
- access to a variegated plant.

Each student should have:

- scissors or scalpel
- forceps
- Bunsen burner, tripod, gauze, heat resistant mat, safety goggles
- boiling tubes, beaker
- white tile, dropping pipette
- ethanol
- iodine solution

Ideally, all the plants should be of the same species. Therefore, it is suggested that variegated and non-variegated geraniums are used. If this is not possible, coleus or variegated tradescantia leaves may be used.

One of the non-variegated plants and the variegated plant should be kept in the light. The other plant should be kept in a dark place for about 24 hours before the assessment takes place.

Safety: teachers will need to ensure that students do not use ethanol near naked flames. This is also stressed in the Student's Instructions.

Procedure
Students will carry out a test for starch on the leaves provided.

Students will be asked to present their results, and to explain them using their scientific knowledge.

Cue sheets
No cue sheets are provided for this assessment task.

Criteria
Using apparatus and equipment

a) methodical preparation of leaf specimens
b) *safe* heating of leaves in ethanol
c) successful application of starch test

Making observations

a) accurate observation of starch test results

Presenting and handling data

a) all results clearly and logically presented

Drawing conclusions

a) correct relationship between light and photosynthesis
b) correct relationship between leaf colour and photosynthesis
c) leaf colour linked to chlorophyll

Background information
There is no background information sheet for this Assessment.

Investigating photosynthesis in leaves

Students' instructions

You need:	scissors or scalpel	forceps
plant with normal leaves kept in light	boiling tube	ethanol
	beaker	white tile
plant with normal leaves kept in dark	Bunsen burner, tripod, gauze	iodine solution
	heat resistant mat	dropping pipette
plant with variegated leaves		

1 Most plants produce their food by *photosynthesis*. One of the products is sugar. Leaves change this into starch so that it can be stored. If we can detect starch in a leaf we can conclude that photosynthesis has been taking place.

2 In this experiment you will test various leaves for starch. You will then use your results to draw some conclusions about photosynthesis.

3 Collect one leaf from each of the plants provided. Make a note of the conditions in which each plant was kept. (Don't get the leaves mixed up!)

4 Cut pieces from each leaf for testing. For the variegated leaf, cut one piece from the green part and one from the 'white' part. The pieces should be large enough to test but small enough to fit in the boiling tube. So that you can identify them, cut the different leaves into different shapes. For example:

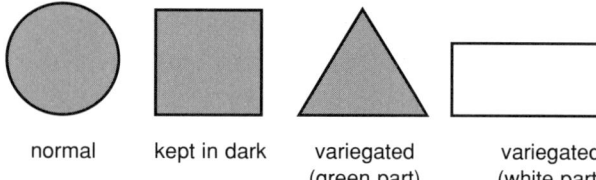

| normal | kept in dark | variegated (green part) | variegated (white part) |

5 *Put on your safety goggles.* Light your Bunsen burner and turn it to a yellow flame. Heat a beaker of water until it is boiling. Dip each piece of leaf in the boiling water for about 30 seconds to kill it.

6 *SAFETY – TURN YOUR BUNSEN OFF.* Leave your beaker of hot water on the tripod.

7 Put all the pieces of leaf into a boiling tube. Add ethanol to the tube so that all the pieces are covered (about half full only). Put the boiling tube into your beaker of hot water. The ethanol will boil and remove much of the leaf colour. This will take about three minutes.

8 Lift the leaf pieces out of the ethanol using forceps. Put them on a white tile. Cover each one with iodine solution. Write down what happens. Record your results in a table. (Remember that a blue-black colour means that starch is present.)

9 What conclusions can you draw from your results? Explain your conclusions using your knowledge of photosynthesis.

This experiment will test your skills in:	**and especially in:**
• using apparatus and equipment	☐
• observing accurately	☐
• presenting results clearly	☐
• drawing conclusions	☐
•	☐

HAZARD WARNING

Ethanol is highly flammable. KEEP AWAY from naked flame. Iodine is harmful to skin and eyes. AVOID SKIN CONTACT. WEAR EYE PROTECTION.

Measuring the energy content of dried bananas

Teacher/technician notes

Skills
This experiment can be used to assess a student's ability in:

- using apparatus and equipment safely
- observing and measuring accurately
- presenting results in a table
- handling data
- evaluating the method used

Previous lessons
Students should be familiar with units of mass and the use of a top-pan balance.

Students will need to be familiar with substituting experimental results into a given formula. Calculators should be available and a cue sheet is provided for those who get stuck.

Apparatus and equipment
Each group should have:

- supply of food sample
- access to top-pan balance
- access to water supply
- calculators

Each student should have:

- Bunsen burner, heat resistant mat
- safety goggles
- tongs for holding food sample
- boiling tube
- measuring cylinder
- retort stand and clamps
- stirring thermometer

The teacher may use any suitable food sample. However, students may have already tried this experiment with peanuts so something unfamiliar may be better. Dried banana slices as sold in supermarkets and health food shops work well. Some are coated with honey but this is not a disadvantage as it aids ignition.

As the banana slices are brittle, they cannot be held on a pin. They can be held in tongs or between the tines of a fork. This may mean that it is difficult to get all of the banana to burn.

Good students will find ways of dealing with this or commenting on it.

Procedure
Students will burn a measured mass of food under a known volume of water. They will measure the temperature rise and hence calculate the energy released.

Students will be asked to present their results, and to discuss the factors affecting the accuracy of their experiments.

Note the use of cue sheets and deduct marks where necessary.

Cue sheets
The cue sheets for this experiment give the following information:

1. sample table for readings
2. sample calculation for calculating energy released
3. sample calculation for calculating energy content (J/g)

Criteria
Using apparatus and equipment

a) correct use of top-pan balance

b) appropriate clamping of boiling tube

c) safety instructions followed

Making observations and measurements

a) mass of food accurately measured

b) volume of water correctly measured

c) starting and final temperatures correctly measured

Presenting and handling data

a) all readings clearly and logically presented in a table

b) calculation successfully made for temperature rise for water, and energy released

Evaluation

a) discussion of possible heat losses

b) awareness of need to repeat measurements

c) suggestions as to how to improve method

Measuring the energy content of dried bananas

Background information

Different foods contain different amounts of energy. This is sometimes known as their *calorific value*. People trying to control their diets need to know the energy content of the foods they are eating. This does not just mean those trying to lose weight – some people, for example, pregnant women, may need to increase their intake of high energy foods.

We can measure the energy content of foods by burning them so that they release energy as heat (thermal energy). We use the following fact to calculate just how much energy has been released.

4.2 joules of energy will raise the temperature of 1 g of water through 1 °C

In this experiment you will measure the energy stored in 1 g of a food sample.

Measuring the energy content of dried bananas	G C S E ASSESSMENT	T3

Students' instructions

You need:

top-pan balance	Bunsen burner, mat, tongs	retort stand and clamps
thermometer	measuring cylinder	boiling tube
safety goggles	calculator	food sample (dried banana)

1 In this experiment you will be measuring the amount of energy released when a food sample burns. Remember to write down any readings you take. Your teacher may wish to check your readings as you take them.

2 Weigh a small slice of dried banana. This is the sample you are going to burn. Write down your result.

3 Measure out 25 ml (cm³) of water and pour it into a boiling tube. Write down the volume of water used. Clamp the boiling tube as shown in the diagram.

4 Take the temperature of the water. Write down your result.

5 Light your Bunsen burner and turn it to a yellow flame.

6 Hold the food sample in the tongs. Turn your Bunsen to a blue flame. Hold the food in the flame until it starts to burn.

7 Hold the burning food under the boiling tube to heat the water. If the food goes out, re-light it immediately and then put it back under the boiling tube. Do this until all the food has burnt.

8 Immediately measure the maximum temperature that the water has reached. (Remember to stir!) Write down your result.

9 You should now have four values in your results table. If you are stuck, ask for **Cue Sheet 1**.

10 Turn your Bunsen to a yellow flame. You may want to repeat your experiment.

11 Calculate the energy released using this equation:

mass of water (g) × temperature rise (°C) × 4.2 (J/g/°C) = energy (J)

Show all your working. If you get stuck, ask for **Cue Sheet 2**.

12 Calculate the energy content of your food sample in joule per gram.

Show all your working. If you get stuck, ask for **Cue Sheet 3**.

13 The manufacturer's value for the energy content of dried banana is 21970 J/g. How does your result compare? Write down any ways in which you think your experiment may have been inaccurate. How could it be improved?

This experiment will test your skills in: and especially in:	HAZARD WARNING
• using apparatus and equipment ☐ • measuring masses accurately ☐ • presenting results in a table ☐ • handling data by doing calculations ☐ • evaluating the method used ☐ • .... ☐	Wear EYE PROTECTION when burning foods.

Measuring the energy content of dried bananas	**G C S E**	**T3**
	ASSESSMENT	

Cue Sheet 1 : setting out the results

Your results should be clearly set out. One possible arrangement is given below:

Mass (weight) of food sample = ... g   (Step 2)

Volume of water used = ... ml   (Step 3)

Starting temperature of water = ... °C   (Step 4)

Final temperature of water = ... °C   (Step 8)

Measuring the energy content of dried bananas	**G C S E**	**T3**
	ASSESSMENT	

Cue Sheet 2 : finding the energy released

To find the amount of energy which went to heating the water you must use the formula given on the instructions sheet. This is an example. Use your results – not these!

Mass (weight) of food sample = 0.7 g   (Step 2)

Volume of water used = 25 ml   (Step 3)

Starting temperature of water = 22 °C   (Step 4)

Final temperature of water = 64 °C   (Step 8)

Formula:

$$\text{mass of water (g)} \times \text{temperature rise (°C)} \times 4.2 \text{ (J/g/°C)} = \text{energy (J)}$$

Now 1 ml of water has a mass of 1 g so we used 25 g of water.

energy = 25 g × (64 − 22) °C × 4.2 J/g/°C

energy = 4410 J

Measuring the energy content of dried bananas	**G C S E**	**T3**
	ASSESSMENT	

Cue Sheet 3 : finding the energy content

To find the energy content of your food sample you can use the following equation:

$$\frac{\textbf{energy released by sample}}{\textbf{mass of food sample}} = \textbf{energy in 1 g of food}$$

So, for example, if 0.7 g of banana releases 4410 J of energy, the energy content is:

$$\frac{4410 \text{ J}}{0.7 \text{ g}} = 6300 \text{ J/g}$$

Testing for glucose

Teacher/technician notes

Skills
This experiment can be used to assess a student's ability in:
- using apparatus and equipment safely
- observing and measuring accurately
- presenting results in a table
- handling data
- drawing conclusions

Previous lessons
Students should be familiar with food tests.

Apparatus and equipment
- Bunsen burner, heat resistant mat, tripod, gauze, safety goggles
- beaker
- five test tubes and rack
- bulb pipette
- Benedict's solution
- access to four glucose solutions of known strength and one unknown

The teacher should prepare five glucose solutions of different strengths from weak to very strong to give a wide range of reactions with Benedict's solution.

The three weakest and the strongest should be labelled with their strengths in grams per litre. The remaining solution should be labelled 'unknown strength'.

Procedure
Students will carry out the test for glucose on the four known solutions. They will note the time taken to observe a change and the final colour of the solution.

From their results they will draw conclusions about the relationship between test results and solution strength.

They will then use their findings to test the unknown solution and hence, estimate its strength.

Note the use of the cue sheet and deduct marks where necessary.

Cue sheets
A cue sheet for this experiment gives a sample table for reading.

Criteria
Using apparatus and equipment

a) safe use of apparatus

Making observations and measurements

a) time for changes accurately measured

b) colours of final solutions well observed

Presenting and handling data

a) all readings and observations for known solutions clearly and logically presented in a table

b) reading and observations clearly presented for unknown solution

Drawing conclusions

a) relationship between time for changes and solution strength clearly stated

b) relationship between colour and solution strength clearly stated

c) strength of unknown solution correctly placed

Background information

There is no background information sheet for this Assessment.

<table>
<tr><td>

Testing for glucose

</td><td>

G C S E
ASSESSMENT

</td><td>

T4

</td></tr>
</table>

Students' instructions

You need:

Bunsen burner, mat, tripod, gauze

beaker, five test tubes, rack

Benedict's solution

dropping pipette

safety goggles

four glucose solutions of known strength

glucose solution of unknown strength

1 In this experiment you will be testing solutions for glucose using Benedict's solution. It is known that the result of this test depends on the strength of the glucose solution.

2 In each of four test tubes, add about 2 cm³ of glucose solution of *known strength*. Label the test tubes so you know which is which!

3 Set up a water bath, as shown in the diagram, in which to heat your solutions. *Put on your safety goggles.* Light the Bunsen and heat the water until it is nearly boiling ('simmering').

4 Add a few drops of Benedict's solution to the weakest of your glucose solutions. Lower the test tube into the hot water. As you do so, start timing.

5 The colour of the solution will change because glucose is present. Time how long it takes the solution to change colour. Write down your result. Also write down the final colour of the solution. (Make sure it has finished changing!)

6 Record your results in a table. If you need help, ask for **Cue Sheet 1**.

7 Repeat the Benedict's test for each of the glucose solutions in turn. Record your results.

8 When you have tested all four known solutions, write down any conclusions which you can draw from your results.

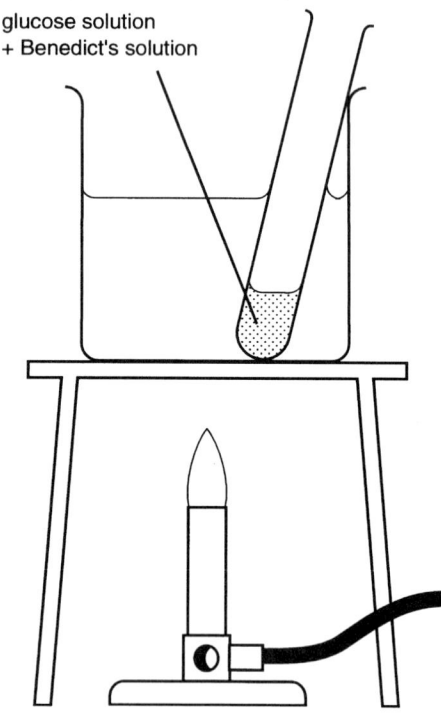

glucose solution
+ Benedict's solution

9 Finally, collect about 2 cm³ of glucose solution of '*unknown strength*' in a test tube. Carry out Benedict's test on this solution. Write down the time taken for it to change colour and describe the final colour.

10 Using your result, estimate the strength of the unknown solution. Write down your answer.

11 Explain how you arrived at your answer.

<table>
<tr><td>

This experiment will test your skills in: **and especially in:**

• using apparatus safely ☐
• measuring times accurately ☐
• presenting results in a table ☐
• drawing conclusions ☐
• explaining conclusions ☐
• .... ☐

</td><td>

HAZARD WARNING

Benedict's is harmful to skin and eyes. AVOID SKIN CONTACT. WEAR EYE PROTECTION.

</td></tr>
</table>

Cue Sheet 1 : setting out the results

Your results should be clearly set out in a table. An example is given below:

solution	strength (in g/l)	time for change (s)	final colour
A			
B			
C			
D			
E			

Measuring the water, humus, and mineral content of soil

Teacher/technician notes

Skills
This experiment can be used to assess a student's ability in:

- using apparatus and equipment safely
- observing and measuring accurately
- presenting results in a table
- presenting results in a graph
- handling data
- evaluating the method used

Previous lessons
Students should be familiar with units of mass and the use of a top-pan balance. They should also be familiar with *safe* methods of heating apparatus to red heat.

Students will need to be familiar with percentage calculations. Calculators should be available and a cue sheet is provided for those who get stuck.

Apparatus and equipment
Each group should have:

- supply of soil
- access to top-pan balance
- calculators.

Each student should have:

- crucible
- Bunsen burner, tripod, pipe-clay
- triangle, heat resistant mat
- safety goggles
- wooden splints.

The soil sample can be taken directly from a garden or the school grounds but some teachers may prefer to prepare a mixture which will give higher proportions of humus and moisture. Such a sample can be prepared by thoroughly mixing garden soil with peat or sieved garden compost. If the materials are very dry, the soil should be *moistened* before use.

Procedure
Students will carry out an experiment to measure the moisture, humus, and mineral content of soil. Masses (weights) will be found using a top-pan balance. Water will be driven off by gently heating. Humus will be burnt off by strong, prolonged heating.

Students will be asked to present their results, calculate percentage proportions, draw a bar chart and to discuss the factors affecting the accuracy of their experiments.

Note the use of cue sheets and deduct marks where appropriate.

Cue sheets
The cue sheets for this experiment give the following information:

1 sample table for readings
2 formula for calculating water content
3 formula for calculating % water content
4 sample bar chart

Criteria
Using apparatus and equipment

a) correct use of top-pan balance
b) good control of Bunsen for gentle/strong heating
c) safety instructions followed

Making observations and measurements

a) masses accurately measured for:

- crucible
- crucible + soil
- crucible + dry soil
- crucible + burnt soil

b) heating stopped at appropriate times

Presenting and handling data

a) all readings clearly and logically presented in a table
b) calculations successfully made for masses of water, humus, minerals
c) calculations successfully made for % of water, humus, minerals
d) bar chart drawn properly

Evaluation

a) discussion of possible soil loss through stirring etc.
b) discussion of difficulty of knowing when all moisture has gone
c) discussion of difficulty of knowing when all humus has burnt away
d) awareness of need to repeat measurements
e) suggestions as to how to improve method

Background information

Different soils are made up of different things. We can show this by shaking up a sample of soil with water. The heavy, insoluble minerals sink to the bottom whilst light, organic matter (humus) floats to the surface.

The properties of the soil depend on the proportions of materials present. For example, a soil with a lot of humus will be rich in nutrients and will hold a lot of water. Soils with lots of large mineral particles will tend to be free-draining. Different plants need different conditions to grow so gardeners and farmers need to know just what sort of soil they have.

In this experiment you will analyse a soil sample. You will measure:

- the amount of water
- the amount of humus
- the amount of minerals

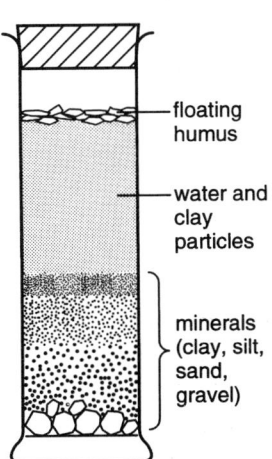

floating humus

water and clay particles

minerals (clay, silt, sand, gravel)

This experiment will test your skills in:	and especially in:
• using apparatus and equipment	☐
• measuring masses (weights) accurately	☐
• presenting results in a table	☐
• handling data by doing calculations	☐
• presenting results as a graph	☐
• evaluating the method used	☐
•	☐

HAZARD WARNING

Take care when heating soil. WEAR EYE PROTECTION. BEWARE SPITTING HOT PARTICLES!

Measuring the water, humus, and mineral content of soil

Students' instructions

You need:

top-pan balance | Bunsen burner, tripod, mat, pipe-clay triangle | crucible

wooden splints | calculator | soil sample

1 In this experiment you will be measuring the amount of water, humus and minerals in a sample of soil. Remember to write down any readings you take. Your teacher may wish to check your readings as you take them.

2 Weigh the empty crucible. Write down your result.

3 Half fill the crucible with soil. Weigh the crucible + soil. Write down your result.

4 Light your Bunsen burner and turn it to a yellow flame. *Put on your safety goggles.*

5 Put your crucible in the pipe-clay triangle and then rest this securely on the tripod.

6 Heat the crucible *very gently* using a small, blue flame. You are going to dry the soil but it *must not burn*. Stir the soil thoroughly with a wooden splint while you are heating it. (Take care not to spill any.) At the first sign of overheating, take the Bunsen away.

7 Keep heating gently until the soil is completely dry.

8 Allow the crucible to cool for at least five minutes. When it is cool enough to touch (be careful!) re-weigh it and write down your result.

9 Put the crucible back on the pipe-clay triangle. *Put on your safety goggles.*

10 Heat the soil *very strongly* until it looks completely burnt.

11 Allow the crucible to cool for at least five minutes. When it is cool enough to touch (be careful!) re-weigh it and write down your result.

12 Your table of results should now contain four masses (weights). If it doesn't, ask for **Cue Sheet 1** which shows a sample table.

13 Use your results to calculate:

- the mass of soil you used

- the mass of water in your soil sample

- the mass of humus in your soil sample

- the mass of miner ils in your soil sample

14 Show all your working. If you get stuck, ask for **Cue Sheet 2**. This shows you how to work out the mass of water.

15 When you have calculated all the masses, calculate the *percentages* of water, humus and minerals in your soil sample. Show all your working. If you get stuck, ask for **Cue Sheet 3**. This shows you how to work out the percentage of water.

16 Draw a bar chart to show the percentage composition of your soil. If you get stuck, ask for **Cue Sheet 4**.

17 Write down any ways in which you think your experiment may have been inaccurate. How could it be improved?

Measuring the water, humus, and mineral content of soil	G C S E		T5
	ASSESSMENT		

Cue Sheet 1 : setting out the results

The results needed to calculate the amounts of water, humus and minerals should be clearly set out. One possible arrangement is given below:

Mass (weight) of empty crucible = ... g (Step 2)

Mass of crucible + soil = ... g (Step 3)

Mass of crucible + dry soil (after gentle heating) = ... g (Step 9)

Mass of crucible + burnt soil (after strong heating) = ... g (Step 11)

- ✂

Measuring the water, humus and mineral content of soil	G C S E		T5
	ASSESSMENT		

Cue Sheet 2 : finding the amount of water

To find the amount of water in your soil sample, you must find out how much mass was lost during the gentle heating:

(mass of crucible + soil) – (mass of crucible + dry soil) = mass of water

(You can use a similar method to find the mass of humus lost during the strong heating. Which results will you need to use?)

- ✂

Measuring the water, humus and mineral content of soil	G C S E		T5
	ASSESSMENT		

Cue Sheet 3 : finding the percentage of water

To find the percentage of water in your soil sample, you can use the following equation:

$$\frac{\text{mass of water in sample}}{\text{mass of soil sample}} \times 100\% = \text{percentage of water}$$

Note that the mass of your soil is given by:

(mass of crucible) – (mass of crucible + soil)

- You can use a similar method to find the percentage of humus. Which results will you need to use?

- How will you find the mineral content? (*Hint:* what should % water + % humus + % mineral add up to?)

Cue Sheet 4 : drawing a bar chart

This is an example only. Use your readings to draw your bar chart.

water content   = 12%

humus content = 21%

mineral content = 67%

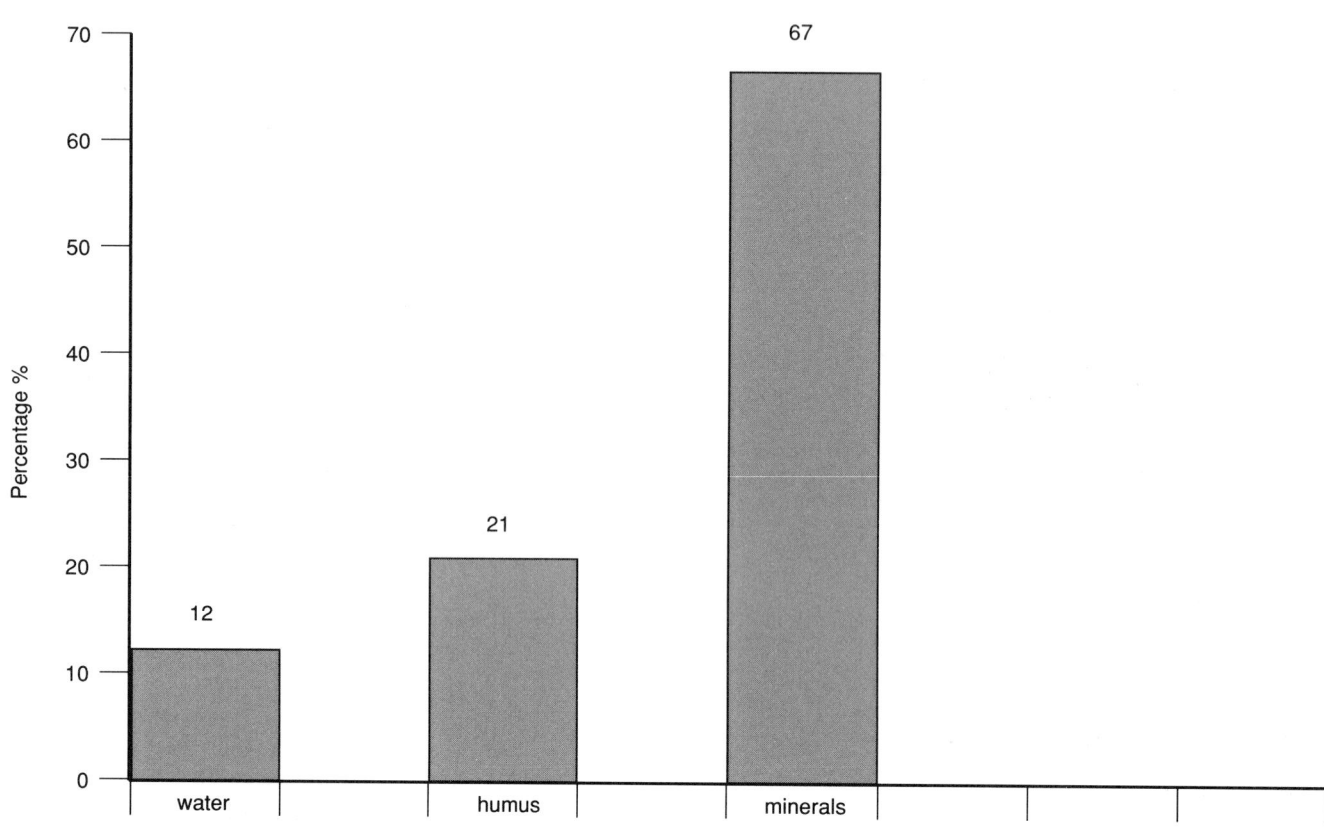

Chart showing the composition of a soil sample

Smoking

Teacher/technician notes

Skills

This assessment can be used to assess a student's ability in:

- making detailed observations
- recording data
- interpreting data from graphs and tables
- constructing a graph from data

Previous lessons

Teachers may find this a useful exercise to include in a course of health education.

Students should be familiar with the structure and workings of lungs.

They should know how to construct a bar graph from data.

Apparatus and materials

The smoking machine, illustrated below, should be set up as a working demonstration so that it is clearly visible to all students. This may be done in a fume cupboard. Smoke inhalation should be avoided.

Each student should have:

- graph and lined paper
- an instruction sheet

Procedure

Students should watch at least three cigarettes being smoked in the demonstration.

They should be encouraged to make detailed observations of what happens in each part of the apparatus.

Written notes of these observations can be made during, or shortly after, the demonstration.

Students then complete the second part of the instruction sheet.

Criteria

Making detailed observations of demonstration

a) colour and density of smoke in 'U' tube
b) colour change of cotton wool
c) colour and density of smoke in first flask
d) colour of water in first flask
e) colour and density of smoke in second flask
f) colour of water in second flask

Interpreting observations of demonstration

a) tar deposits could form in the lungs
b) chemicals from smoke will dissolve in moisture in the lungs

Answers to question 1 – interpretation of bar graphs

a) 10 in 100 000
b) 110 in 100 000
c) eleven times more likely
d) yes
e) yes – all ages have increased risk
f) 11 in 100 000
g) 110 in 100 000
h) ten times more likely
i) two times and two-and-a-half times more likely

Answer to question 2 – construct graph from data

a) title
b) axes labelled
c) plotting accurate
d) neatness

Interpreting graph

The more cigarettes you smoke, the greater the risk of dying from cancer.

Diagram of smoking machine

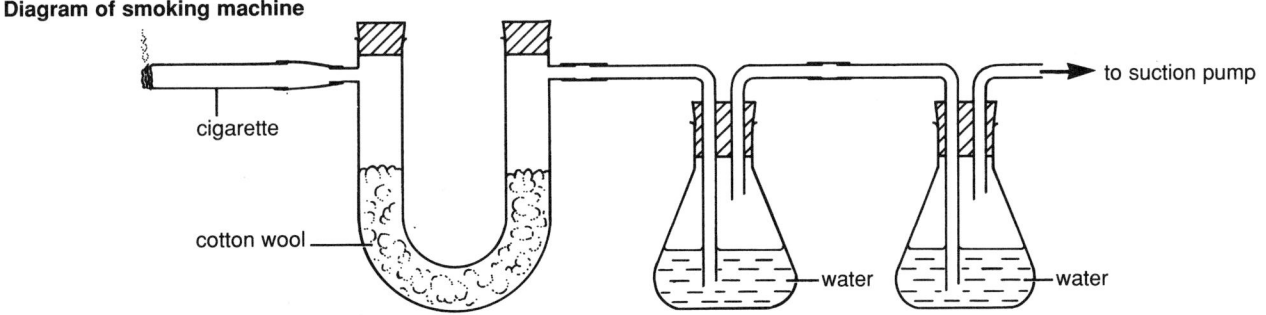

<table>
<tr><td></td><td>G C S E</td><td rowspan="2">T6</td></tr>
<tr><td># Smoking</td><td>ASSESSMENT</td></tr>
</table>

Students' Instructions – part A

1 Make detailed observations of the smoking machine demonstration provided.

2 You will be assessed on your ability to:
   - *describe what happens to the smoke in each part of the apparatus*
   - *describe all the effects which smoke has on parts of the apparatus*
   - *describe what you think might happen inside a smoker's lungs while a cigarette is smoked.*

Students' Instructions – part B

1 Study the two graphs below.

2 *You will be assessed on your ability to interpret the graphs by answering the questions that follow as fully as possible.*

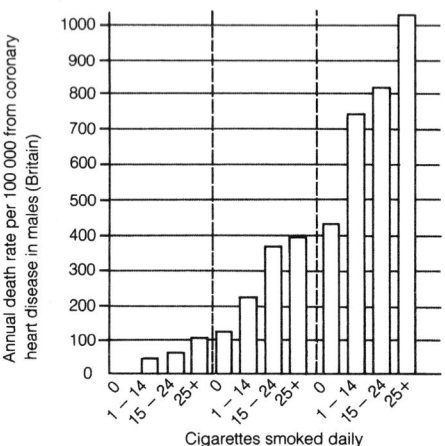 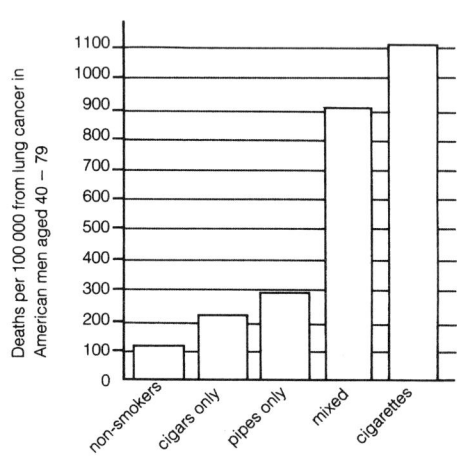

a) What is the risk of a non-smoker having a heart attack under the age of 45?

b) What is the risk of having a heart attack under the age of 45 if you smoke 25 cigarettes a day?

c) How many times more likely are you to have a heart attack under the age of 45 if you smoke 25 cigarettes a day?

d) Are smokers more likely to die of heart disease, at any age, than non-smokers?

e) Is there any evidence that the number of cigarettes you smoke affects the risk of having a heart attack?

f) What is the risk of dying of lung cancer if you are a non-smoker?

g) What is the risk of dying of lung cancer if you smoke cigarettes?

h) How many times more likely are you to die of lung cancer if you smoke?

i) What health risks are involved with smoking pipes and cigars?

3 Study the table below:

4 *You will be assessed on your ability to present the information in this table as a bar graph.*

5 What does this graph tell you about health risks from smoking cigarettes?

Cigarette smoking and lung cancer	number of cigarettes per day	deaths from lung cancer
	none	12 per 100 000
	1 to 14	96 per 100 000
	15 to 24	156 per 100 000
	more than 25	300 per 100 000